Dimensions Math®
Teacher's Guide 3A

Authors and Reviewers

Cassandra Turner

Allison Coates

Jenny Kempe

Bill Jackson

Tricia Salerno

Singapore Math Inc.

Published by Singapore Math Inc.

19535 SW 129th Avenue
Tualatin, OR 97062
www.singaporemath.com

Dimensions Math® Teacher's Guide 3A
ISBN 978-1-947226-36-4

First published 2018
Reprinted 2019, 2020 (twice)

Printed in China

Acknowledgments

Editing by the Singapore Math Inc. team.
Design and illustration by Cameron Wray with Carli Bartlett.

Contents

Chapter		Lesson	Page

Chapter		Lesson	Page

Chapter		Lesson	Page

Chapter		Lesson	Page

Dimensions Math® Curriculum

The **Dimensions Math®** series is a Pre-Kindergarten to Grade 5 series based on the pedagogy and methodology of math education in Singapore. The main goal of the **Dimensions Math®** series is to help students develop competence and confidence in mathematics.

The series follows the principles outlined in the Singapore Mathematics Framework below.

Pedagogical Approach and Methodology

- Through Concrete-Pictorial-Abstract development, students view the same concepts over time with increasing levels of abstraction.
- Thoughtful sequencing creates a sense of continuity. The content of each grade level builds on that of preceding grade levels. Similarly, lessons build on previous lessons within each grade.
- Group discussion of solution methods encourages expansive thinking.
- Interesting problems and activities provide varied opportunities to explore and apply skills.
- Hands-on tasks and sharing establish a culture of collaboration.
- Extra practice and extension activities encourage students to persevere through challenging problems.
- Variation in pictorial representation (number bonds, bar models, etc.) and concrete representation (straws, linking cubes, base ten blocks, discs, etc.) broaden student understanding.

Each topic is introduced, then thoughtfully developed through the use of a variety of learning experiences, problem solving, student discourse, and opportunities for mastery of skills. This combination of hands-on practice, in-depth exploration of topics, and mathematical variability in teaching methodology allows students to truly master mathematical concepts.

Singapore Mathematics Framework

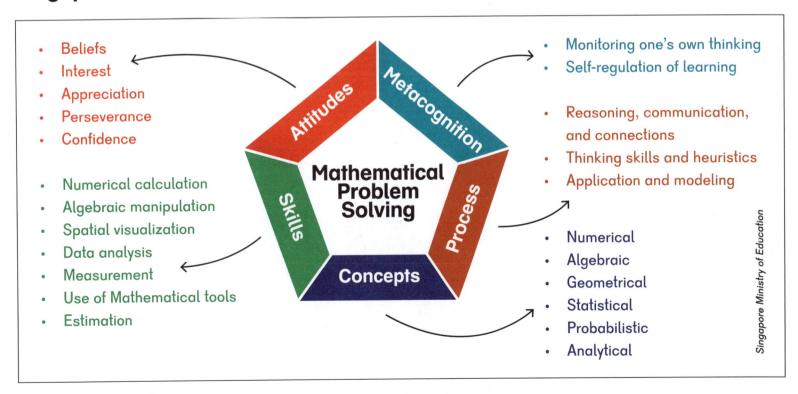

Singapore Ministry of Education

Dimensions Math® Program Materials

Textbooks

Textbooks are designed to help students build a solid foundation in mathematical thinking and efficient problem solving. Careful sequencing of topics, well-chosen problems, and simple graphics foster deep conceptual understanding and confidence. Mental math, problem solving, and correct computation are given balanced attention in all grades. As skills are mastered, students move to increasingly sophisticated concepts within and across grade levels.

Students work through the textbook lessons with the help of five friends: Emma, Alex, Sofia, Dion, and Mei. The characters appear throughout the series and help students develop metacognitive reasoning through questions, hints, and ideas.

A pencil icon ✏ at the end of the textbook lessons links to exercises in the workbooks.

Workbooks

Workbooks provide additional problems that range from basic to challenging. These allow students to independently review and practice the skills they have learned.

Teacher's Guides

Teacher's Guides include lesson plans, mathematical background, games, helpful suggestions, and comprehensive resources for daily lessons.

Tests

Tests contain differentiated assessments to systematically evaluate student progress.

Emma   Alex   Sofia   Dion   Mei

Online Resources

The following can be downloaded from dimensionsmath.com.

- **Blackline Masters** used for various hands-on tasks.

- **Material Lists** for each chapter and lesson, so teachers and classroom helpers can prepare ahead of time.

- **Activities** that can done with students who need more practice or a greater challenge, organized by concept, chapter, and lesson.

- **Standards Alignments** for various states.

Using the Teacher's Guide

This guide is designed to assist in planning daily lessons. It should be considered a helping hand between the curriculum and the classroom. It provides introductory notes on mathematical content, key points, and suggestions for activities. It also includes ideas for differentiation within each lesson, and answers and solutions to textbook and workbook problems.

Each chapter of the guide begins with the following.

● Overview

Includes objectives and suggested number of class periods for each chapter.

● Notes

Highlights key learning points, provides background on math concepts, explains the purpose of certain activities, and helps teachers understand the flow of topics throughout the year.

● Materials

Lists materials, manipulatives, and Blackline Masters used in the Think and Learn sections of the guide. It also includes suggested storybooks. Many common classroom manipulatives are used throughout the curriculum. When a lesson refers to a whiteboard and markers, any writing materials can be used. Blackline Masters can be found at dimensionsmath.com.

The guide goes through the Chapter Openers, Daily Lessons, and Practices of each chapter, and cumulative reviews in the following general format.

● Chapter Opener

Provides talking points for discussion to prepare students for the math concepts to be introduced.

● Think

Offers structure for teachers to guide student inquiry. Provides various methods and activities to solve initial textbook problems or tasks.

● Learn

Guides teachers to analyze student methods from Think to arrive at the main concepts of the lesson through discussion and study of the pictorial representations in the textbook.

● Do

Expands on specific problems with strategies, additional practice, and remediation.

● Activities

Allows students to practice concepts through individual, small group, and whole group hands-on tasks and games, including suggestions for outdoor play (most of which can be modified for a gymnasium or classroom).

Level of difficulty in the games and activities are denoted by the following symbols.

- ● Foundational activities
- ▲ On-level activities
- ★ Challenge or extension activities

● Brain Works

Provides opportunities for students to extend their mathematical thinking.

Discussion is a critical component of each lesson. Teachers are encouraged to let students discuss their reasoning. As each classroom is different, this guide does not anticipate all situations. The following questions can help students articulate their thinking and increase their mastery:

- Why? How do you know?
- Can you explain that?
- Can you draw a picture of that?
- Is your answer reasonable? How do you know?
- How is this task like the one we did before? How is it different?
- What is alike and what is different about…?
- Can you solve that a different way?
- Yes! You're right! How do you know it's true?
- What did you learn before that can help you solve this problem?
- Can you summarize what your classmate shared?
- What conclusion can you draw from the data?

Each lesson is designed to take one day. If your calendar allows, you may choose to spend more than one day on certain lessons. Throughout the guide, there are notes to extend on learning activities to make them more challenging. Lesson structures and activities do not have to conform exactly to what is shown in the guide. Teachers are encouraged to exercise their discretion in using this material in a way that best suits their classes.

Textbooks are designed to last multiple years. Textbook problems with a ▨ (or a blank line for terms) are meant to invite active participation.

Dimensions Math® Scope & Sequence

PKA

Chapter 1
Match, Sort, and Classify

Red and Blue
Yellow and Green
Color Review
Soft and Hard
Rough, Bumpy, and Smooth
Sticky and Grainy
Size — Part 1
Size — Part 2
Sort Into Two Groups
Practice

Chapter 2
Compare Objects

Big and Small
Long and Short
Tall and Short
Heavy and Light
Practice

Chapter 3
Patterns

Movement Patterns
Sound Patterns
Create Patterns
Practice

Chapter 4
Numbers to 5 — Part 1

Count 1 to 5 — Part 1
Count 1 to 5 — Part 2
Count Back

Count On and Back
Count 1 Object
Count 2 Objects
Count Up to 3 Objects
Count Up to 4 Objects
Count Up to 5 Objects
How Many? — Part 1
How Many? — Part 2
How Many Now? — Part 1
How Many Now? — Part 2
Practice

Chapter 5
Numbers to 5 — Part 2

1, 2, 3
1, 2, 3, 4, 5 — Part 1
1, 2, 3, 4, 5 — Part 2
How Many? — Part 1
How Many? — Part 2
How Many Do You See?
How Many Do You See Now?
Practice

Chapter 6
Numbers to 10 — Part 1

0
Count to 10 — Part 1
Count to 10 — Part 2
Count Back
Order Numbers
Count Up to 6 Objects
Count Up to 7 Objects
Count Up to 8 Objects
Count Up to 9 Objects
Count Up to 10 Objects
— Part 1

Count Up to 10 Objects
— Part 2
How Many?
Practice

Chapter 7
Numbers to 10 — Part 2

6
7
8
9
10
0 to 10
Count and Match — Part 1
Count and Match — Part 2
Practice

PKB

Chapter 8
Ordinal Numbers

First
Second and Third
Fourth and Fifth
Practice

Chapter 9
Shapes and Solids

Cubes, Cylinders, and Spheres
Cubes
Positions
Build with Solids
Rectangles and Circles
Squares
Triangles

Dimensions Math® Scope & Sequence

Count Up to 10 Things —
    Part 2
Recognize the Numbers
    6 to 10
Write the Numbers 6 and 7
Write the Numbers 8, 9,
    and 10
Write the Numbers 6 to 10
Count and Write the
    Numbers 1 to 10
Ordinal Positions
One More Than
Practice

Chapter 4
Shapes and Solids

Curved or Flat
Solid Shapes
Closed Shapes
Rectangles
Squares
Circles and Triangles
Where is It?
Hexagons
Sizes and Shapes
Combine Shapes
Graphs
Practice

Chapter 5
Compare Height, Length, Weight, and Capacity

Comparing Height
Comparing Length
Height and Length — Part 1
Height and Length — Part 2
Weight — Part 1

Weight — Part 2
Weight — Part 3
Capacity — Part 1
Capacity — Part 2
Practice

Chapter 6
Comparing Numbers Within 10

Same and More
More and Fewer
More and Less
Practice — Part 1
Practice — Part 2

KB

Chapter 7
Numbers to 20

Ten and Some More
Count Ten and Some More
Two Ways to Count
Numbers 16 to 20
Number Words 0 to 10
Number Words 11 to 15
Number Words 16 to 20
Number Order
1 More Than or Less Than
Practice — Part 1
Practice — Part 2

Chapter 8
Number Bonds

Putting Numbers Together
    — Part 1

Putting Numbers Together
    — Part 2
Parts Making a Whole
Look for a Part
Number Bonds for 2, 3, and 4
Number Bonds for 5
Number Bonds for 6
Number Bonds for 7
Number Bonds for 8
Number Bonds for 9
Number Bonds for 10
Practice — Part 1
Practice — Part 2
Practice — Part 3

Chapter 9
Addition

Introduction to Addition —
    Part 1
Introduction to Addition —
    Part 2
Introduction to Addition —
    Part 3
Addition
Count On — Part 1
Count On — Part 2
Add Up to 3 and 4
Add Up to 5 and 6
Add Up to 7 and 8
Add Up to 9 and 10
Addition Practice
Practice

Chapter 10
Subtraction

Take Away to Subtract —
    Part 1

Dimensions Math® Scope & Sequence

Dimensions Math® Scope & Sequence

Dimensions Math® Scope & Sequence

Dimensions Math® Scope & Sequence

Dimensions Math® Scope & Sequence

Suggested number of class periods: 11–12

	Lesson	Page	Resources		Objectives
	Chapter Opener	p. 5	TB:	p. 1	Investigate numbers to 10,000.
1	Numbers to 10,000	p. 6	TB: WB:	p. 2 p. 1	Understand that a four-digit number is composed of thousands, hundreds, tens, and ones.
2	Place Value — Part 1	p. 10	TB: WB:	p. 8 p. 4	Identify the value and the place of each digit in a four-digit number.
3	Place Value — Part 2	p. 12	TB: WB:	p. 12 p. 7	Determine the number of tens in a four-digit number that is a multiple of 10 and the number of hundreds in a four-digit number that is a multiple of 100.
4	Comparing Numbers	p. 15	TB: WB:	p. 16 p. 10	Compare numbers to 10,000.
5	The Number Line	p. 18	TB: WB:	p. 20 p. 13	Locate numbers on a number line.
6	Practice A	p. 21	TB: WB:	p. 24 p. 18	Practice concepts from the chapter.
7	Number Patterns	p. 23	TB: WB:	p. 27 p. 22	Find a number that is 10, 100, or 1,000 more or less than a four-digit number.
8	Rounding to the Nearest Thousand	p. 26	TB: WB:	p. 31 p. 25	Round a four-digit number to the nearest thousand.
9	Rounding to the Nearest Hundred	p. 28	TB: WB:	p. 33 p. 27	Round a four-digit number to the nearest hundred.
10	Rounding to the Nearest Ten	p. 30	TB: WB:	p. 35 p. 30	Round a four-digit number to the nearest ten.
11	Practice B	p. 32	TB: WB:	p. 38 p. 33	Practice skills and concepts from the chapter.
	Workbook Solutions	p. 34			

In **Dimensions Math® 2**, students learned to:

- Relate three-digit numbers to place value.
- Use base ten blocks and place-value discs to show a three-digit number.
- Use place-value charts to form numbers.
- Compare 2 three-digit numbers.

In this chapter, students build upon that knowledge and extend concepts and skills to four-digit numbers.

The position of the digit in relation to other digits determines its value. Each place represents a value ten times the place to its right.

The number 2,368, for example, equals 2 thousands, 3 hundreds, 6 tens, 8 ones.

The digit 3 is in the hundreds place. Its value is 300.

The number 2,368 can also be written as a sum of its place values, or in expanded form:
2,000 + 300 + 60 + 8

Place-value Discs

In order to gain a solid understanding of place value, students should have sufficient hands-on experience with manipulatives and see many different representations of place value.

Students should be familiar with base ten blocks from **Dimensions Math® 2**, however, thousands, ten thousands, and larger numbers are impractical to represent with base ten blocks.

In **Dimensions Math® 2**, students used place-value discs. Place-value discs are more abstract than base ten blocks as they are not proportional in size to the numbers they represent.

Students will continue to use place-value discs in lessons in Dimensions Math 4 and 5, when they will be used to introduce place value in decimal numbers.

Students have learned that a 10-disc represents the value of 10 ones and a 100-disc represents the value of 10 tens. In **Dimensions Math® 3**, students will extend their knowledge to the thousands place using 1,000-discs.

Once students transition to place-value discs, base ten blocks and other ways to represent numbers will still be used periodically in the textbook when introducing new concepts.

Students who continue to struggle with knowing that a 10-disc represents 10 units should use base ten blocks.

Place-value Charts and Organizers

In **Dimensions Math® 3A**, students will use a place-value chart with columns for thousands, hundreds, tens, and ones. Students write numbers in the correct column according to its place.

Using place-value manipulatives on a place-value chart can be confusing at first. Some students may think, for example, that a 10-disc in the tens column is worth 10 tens instead of 1 ten. Using a place-value organizer, which has headers to indicate place value, will help prevent this confusion.

Thousands	Hundreds	Tens	Ones
2	3	6	8

If discs are used, there are no headers indicating which column is hundreds, tens, or ones.

Students should have a way of organizing their discs by place to differentiate them from other discs on their desks. They can do this by simply creating 4 distinct areas on their desks, or by using a paper mat divided into columns.

Each student, or pair of students, will need twenty each of 100, 10, and 1-discs, and ten 1,000-discs for this chapter.

Place-value Cards (also called Place-value Strips)

Place-value cards are provided as a Blackline Master. Students should have their own sets for numbers to the thousands. Place-value cards help students understand that numbers can be composed or decomposed according to the value of each digit.

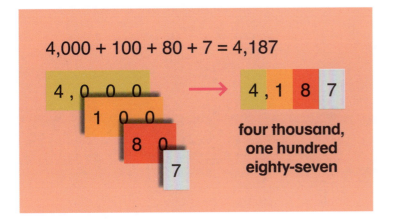

Number Lines and Rounding

Rounding is a specific skill that is used for estimating calculations. This chapter introduces number lines for rounding numbers. Number lines help students visualize the position of a number relative to the nearest ten or multiple of ten when rounding.

Students will learn to use number lines that do not show 0 as the starting point. They will also be introduced to number lines in which the increment between two tick marks is not 1. These concepts may be difficult at first.

Students begin by rounding to the nearest thousand. 2,368 is nearer to 2,000 than 3,000, and is rounded to 2,000.

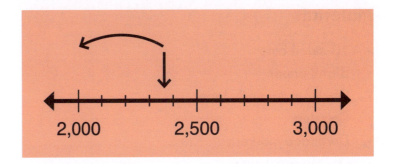

Students often have difficulty locating the position of a number on the number line. In the example of 2,368, rounding to the nearest ten requires students to identify the nearest multiples of ten as 2,360 and 2,370.

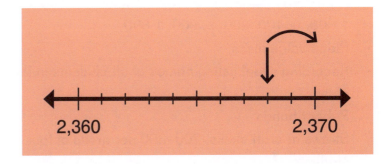

By convention, numbers that are halfway between, or equally far from, the two multiples are often rounded to the greater number. The number 2,365 rounded to the nearest ten is 2,370.

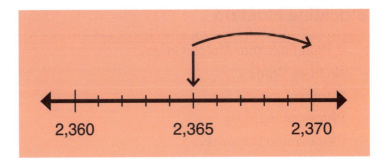

Students will learn which place they can use for rounding. For example, when rounding a number to the nearest hundred, students will look at the number in the tens place. In 2,365, the tens place digit is 6, which rounds to the next hundred, or 2,400.

Materials

- 10-sided dice
- Blank paper
- Chalk or painter's tape
- Counters
- Dry erase markers
- Dry erase sleeves
- Name tags with decomposed numbers written on them
- Note cards, folded in half with multiples of 10 to 100: 0, 10, 20, 30, ... 90, 100
- Note cards, folded in half with multiples of 100 to 1,000: 0, 100, 200, ... 900, 1,000
- Place-value discs
- Recording sheet listing names of all students in the class
- Rubber bands
- Straws or craft sticks, 200–300 per group of four students
- String or yarn, 12 feet
- Two-color counters
- Whiteboards

Blackline Masters

- 4-Digit Find Your Match Cards
- Number Cards
- Number Line
- Place-value Cards
- Place-value Organizer
- Shut the Box — Thousands Game Board
- Three in a Row — Rounding to Hundreds Game Board

Storybooks

- *Place Value* by David A. Adler
- *Sir Cumference and All the King's Tens: A Math Adventure* by Cindy Neuschwander
- *How Big is a Million* by Anna Milbourn
- *Earth Day — Hooray* by Stuart J. Murphy
- *Really Big Numbers* by Richard Evan Schwartz

Activities

Games and activities included in this chapter are designed to provide practice and extensions of place value concepts. They can be used after students complete the **Do** questions, or any time review and practice are needed.

Chapter Opener

Objective

- Investigate numbers to 10,000.

Lesson Materials

- Straws or craft sticks, 200–300 per group of four students
- Rubber bands

Provide groups of four students with 200–300 straws or craft sticks. Ask them to guess how many they have and discuss how to count them.

Have each group bundle their straws into groups of 10 ones, then bundle 10 tens to make a hundred.

Have groups combine straws to determine how many the whole class has. Ask students:

- What happens to number names when we bundle groups of ten? (10 ones is equal to 1 ten, 10 tens is equal to 1 hundred.)
- Do you know what the name for 10 hundreds is? (10 hundreds is equal to 1 thousand.)

Save the bundled straws and continue to Lesson 1.

Chapter 1

Numbers to 10,000

How can we count all of these straws easily without losing track of how many we have counted?

1

Lesson 1 Numbers to 10,000

Objective

- Understand that a four-digit number is composed of thousands, hundreds, tens, and ones.

Lesson Materials

- Place-value Cards (BLM)
- Bundled straws from **Chapter Opener**
- Place-value discs

Think

Pose the **Think** problem and have pairs of students represent the number of straws or craft sticks from the **Chapter Opener** with Place-value Cards (BLM) and place-value discs.

Discuss the strategies students used to find their answers to the **Think** problem.

Examples:

- I counted the thousands first: 2,000, then 2 thousand 4 hundred, then 2 thousand 4 hundred thirty, then 2 thousand 4 hundred thirty-six.
- I counted each value and found 2 thousands plus 4 hundreds plus 3 tens plus 6 ones.

Students may find alternative strategies.

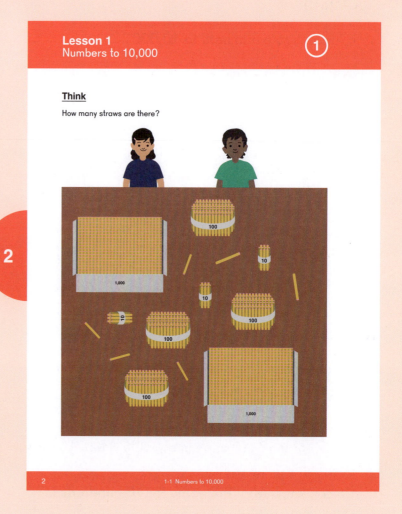

Lesson 1
Numbers to 10,000 ①

Think

How many straws are there?

2   1-1 Numbers to 10,000

Learn

Students should see from the use of Place-value Cards (BLM) that the written numeral shows the number of thousands, hundreds, tens, and ones from left to right.

Have students discuss the different representations of the number 2,436:

- Base ten blocks
- Place-value Cards (BLM)
- Number words
- Expanded notation

Discuss Emma's comment about the comma in the thousands numbers. This is a convention to make the number easier to read.

Students will continue showing and writing four-digit numbers throughout the extended **Do** portion of this lesson.

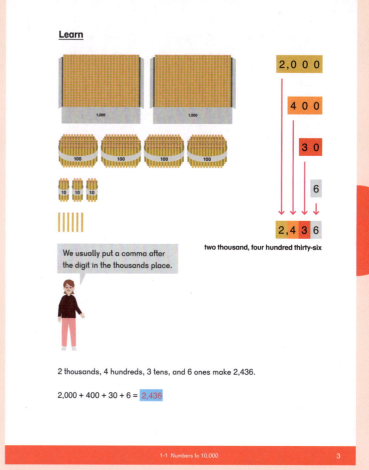

Do

1 Discuss the different representations of each number. Most students should be working with just the pages in the textbook.

(b) There are no ones in the problem. Students should note that there is no need to write 1,000 + 300 + 20 + 0, however, we do represent 0 ones in the written number 1,320.

(c) There is no need to write 2,000 + 100 + 00 + 9, however, we do represent 0 tens in the written number 2,109.

Ask students what would happen if we didn't represent 0 tens in the written number.

(d) There is no need to write 4,000 + 000 + 70 + 6, however, we do represent 0 hundreds in the written number 4,076 because 00 and 000 are not numbers.

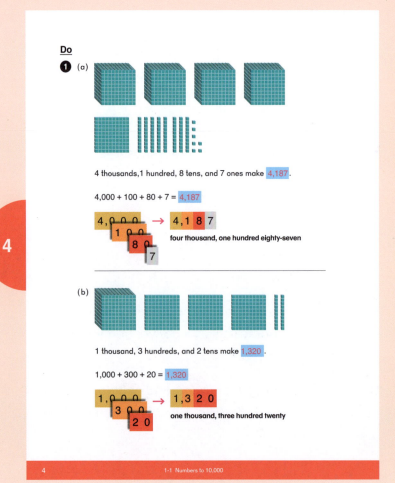

❷ This problem introduces the number 10,000. Have students trade ten 1,000-discs for one 10,000-disc.

❹ — ❻ Have students show the numbers on whiteboards to assess for understanding.

Activities

▲ Match Me

Materials: Place-value Cards (BLM)

One partner creates a four-digit number with Place-value Cards (BLM). His partner writes the number in expanded form. If a third person is playing, she can write the number in numerals or number words.

▲ Place-value Hop

Materials: Chalk or painter's tape

Using chalk on the sidewalk, or painter's tape indoors, make a 4-row place-value board.

The Hopper picks a four-digit number and hops his number. His partner needs to figure out the number.

1,000
100
10
1

For the number 4,605, the Hopper hops:
- Four times on the 1,000
- Six times on the 100
- Over the 10
- Five times on the 1

For a greater challenge, have the Hopper hop the places out of order:
- Six times on the 100
- Five times on the 1
- Over the 10
- Four times on the 1,000

Exercise 1 • page 1

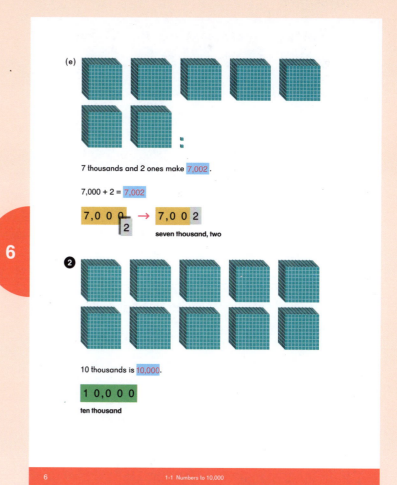

(e)

7 thousands and 2 ones make 7,002.

7,000 + 2 = 7,002

7,0 0 0 / 2 → 7,0 0 2
seven thousand, two

❷

10 thousands is 10,000.

1 0,0 0 0
ten thousand

❸ Read each number.
Show each number with place-value cards.
Write each number in words.

(a) 1,299   (b) 9,494   (c) 7,250
one thousand, two hundred ninety-nine   nine thousand, four hundred ninety-four   seven thousand, two hundred fifty
(d) 2,306   (e) 5,027   (f) 6,005
two thousand, three hundred six   five thousand, twenty-seven   six thousand five

❹ (a) 3,000 + 300 + 40 + 9 = 3,349

(b) 7,000 + 80 + 1 = 7,081

(c) 9,000 + 600 + 4 = 9,604

(d) 5,000 + 3 = 5,003

(e) 5 + 90 + 400 + 6,000 = 6,495

❺ Count on by ones from 4,987 to 5,011. 4,987, 4,988, 4,989, 4,990, 4,991, 4,992, 4,993, 4,994, 4,995, 4,996, 4,997, 4,998, 4,999, 5,000, 5,001, 5,002, 5,003, 5,004, 5,005, 5,006, 5,007, 5,008, 5,009, 5,010, 5,011

❻ Count on by ones and write the missing numbers.

(a) 7,007   7,008   7,009   7,010   7,011   7,012   7,013

(b) 6,996   6,997   6,998   6,999   7,000   7,001   7,002

Exercise 1 • page 1

Lesson 2 Place Value —Part 1

Objective

- Identify the value and the place of each digit in a four-digit number.

Lesson Materials

- Place-value Cards (BLM)
- Place-value discs
- Place-value Organizer (BLM)

Think

Provide students with place-value discs and a Place-value Organizer (BLM). Have them try to solve the **Think** problem independently.

Learn

Discuss the **Learn** examples. Ask students:

- What is meant by the "value" of each digit?
- What does a digit stand for, and what is its value?

Students will continue working with digits and their values throughout the **Do** portion of this lesson. In addition to asking how many tens there are, ask, "What is the value of the digit in the tens place?"

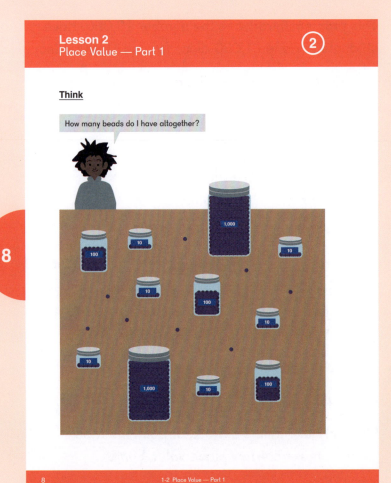

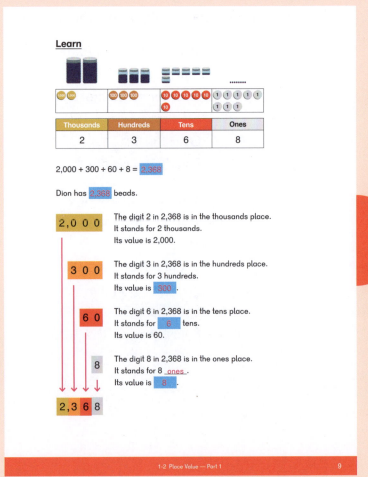

Do

If needed, provide additional practice with different four-digit numbers and questions similar to ❷ and ❺, where the values given are not necessarily in order from greatest to least.

❺ It is important that students are able to compose the number when the numbers being added are not in order from greatest to least.

If students are confused, give them Place-value Cards (BLM) to make the number and think about what is missing.

Activity

▲ Place-value Hangman

Students play hangman using four-digit numbers.

Player One makes a four-digit number and draws 4 lines:

_____ , _____ _____ _____

Player Two tries to guess the number by asking yes/no questions like:

- Is there a 3 in the tens place?
- Is the digit in the hundreds place greater than 5?
- Is the value of the thousands digit less than 4?

Exercise 2 • page 4

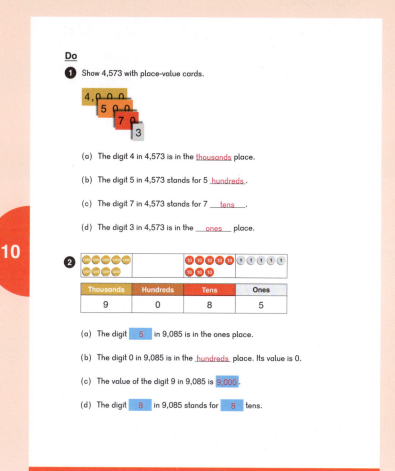

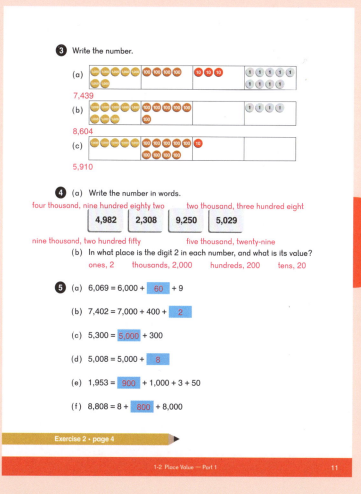

Objective

- Determine the number of tens in a four-digit number that is a multiple of 10 and the number of hundreds in a four-digit number that is a multiple of 100.

Lesson Materials

- Place-value discs

Think

Provide students with place-value discs and adequate time to work through the **Think** problem.

Have students share their strategies for solving Dion and Emma's problems.

Examples:

- I made the numbers with just 100-discs (or 10-discs).
- I thought of 2,300 as 23 hundred, so I used twenty-three 100-discs.
- I knew 230 was 23 tens, so I thought 2,300 was 23 tens times ten.
- I counted the zeros.

Students may find alternative strategies.

Learn

Discuss the progression of examples in **Learn**.
In (b), Dion uses 100-discs:

1,000 = 10 hundreds
2,000 = 2 × 10 hundreds = 20 hundreds
300 = 3 hundreds
2,300 = 20 hundreds + 3 hundreds = 23 hundreds

Alex notes that the number 2,3**00** is the same value as 23 **hundreds**.

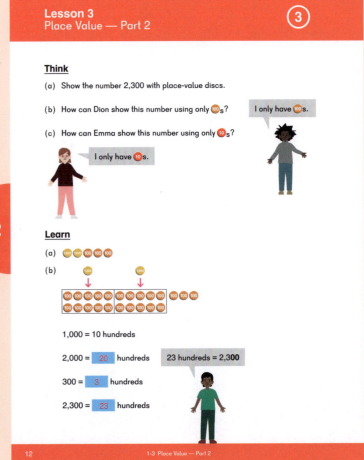

In (c), Emma uses 10-discs:

1,000 = 100 tens
2,000 = 2 × 100 tens = 200 tens
300 = 30 tens
2,300 = 200 tens + 30 tens = 230 tens

Mei notes that the number 2,300 is the same value as 230 tens.

Note: 230 tens is hard to show with place-value discs.

The question, "How many tens are in 1,000?" is not asking for the value of the digit in the tens place. It is asking, instead, "If 1 thousand is regrouped as tens, how many tens would there be?"

Do

Discuss the examples with students.

❶ — ❷ Point out that four-digit numbers are often expressed as hundreds but not in tens. It is common to say "67 hundred" for 6,700.

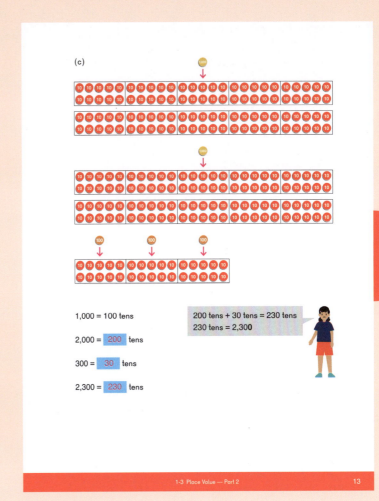

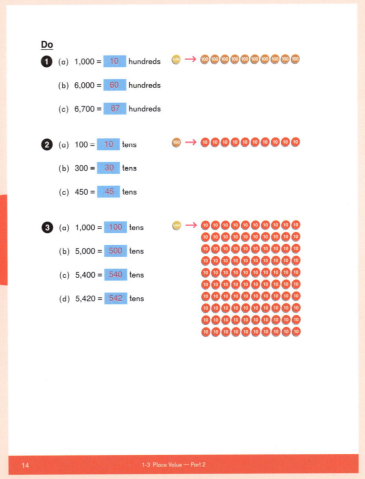

4 Have students write equivalent expressions.

(a) 4,900 = 49 hundreds

(b) 7,200 = 72 hundreds

(c) 5,600 = 560 tens

(d) 9,210 = 921 tens

These problems can also be written as the sum of place values:

(a) 40 hundreds + 9 hundreds = 49 hundreds

(b) 70 hundreds + 2 hundreds = 72 hundreds

(c) 500 tens + 60 tens = 560 tens

(d) 920 tens + 1 ten = 921 tens

6 (b) This problem shows that numbers can be expressed in different ways: 4 thousands + 32 tens instead of 432 tens as in previous examples.

(c) 2,047 can also be expressed as:
- 2 thousands, 47 ones
- 20 hundreds, 4 tens, 7 ones
- 2,047 ones

To extend, ask, "How many ways can you express the number…?"

Activity

▲ **Find Your Match — 4-Digit**

Materials: 4-Digit Find Your Match Cards (BLM)

This can be a whole class activity that uses number words, numbers in expanded form, numerals, and place-value chart representations.

Pass out 4-Digit Find Your Match Cards (BLM) to students and have students find cards with numbers or representations that match their own cards.

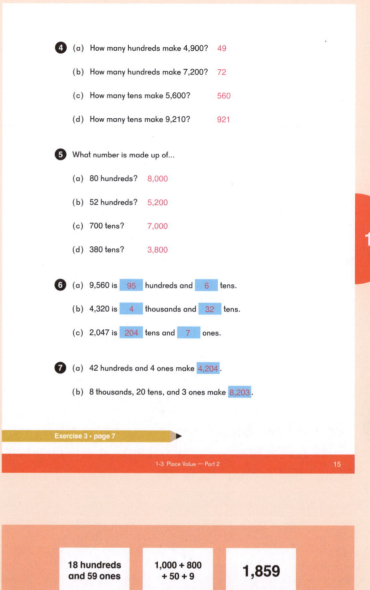

4 (a) How many hundreds make 4,900?   49

(b) How many hundreds make 7,200?   72

(c) How many tens make 5,600?   560

(d) How many tens make 9,210?   921

5 What number is made up of…

(a) 80 hundreds?   8,000

(b) 52 hundreds?   5,200

(c) 700 tens?   7,000

(d) 380 tens?   3,800

6 (a) 9,560 is [95] hundreds and [6] tens.

(b) 4,320 is [4] thousands and [32] tens.

(c) 2,047 is [204] tens and [7] ones.

7 (a) 42 hundreds and 4 ones make [4,204].

(b) 8 thousands, 20 tens, and 3 ones make [8,203].

Exercise 3 • page 7

1-3 Place Value — Part 2   15

15

| 18 hundreds and 59 ones | 1,000 + 800 + 50 + 9 | 1,859 |

Exercise 3 • page 7

Lesson 4 Comparing Numbers

Objective

- Compare numbers to 10,000.

Lesson Materials

- Place-value Cards (BLM)
- Place-value discs

Think

Provide students with place-value discs and Place-value Cards (BLM) and pose the **Think** problems. Ask what strategies they could use to figure out which game had the most attendees and which game had the least.

Discuss the different strategies. Examples:

- 381 tens is more than 317 tens so the soccer match had the most people.
- I looked at the digits in each place value.
- I made the numbers with discs and counted which had more.

Students may find alternative strategies.

Learn

Discuss Dion's comment. Have students compare their solutions from **Think** with the one shown in the textbook.

Guide students to look first at the digit in the thousands place to compare four-digit numbers, then look at the digit in the hundreds place.

Review the meaning of the "greater than" and "less than" signs.

Note: If students have trouble remembering which way the sign should point, have them make two dots next to the greater number and one dot next to the lesser number. When students connect the dots, they create the correct sign.

Lesson 4
Comparing Numbers ④

Think

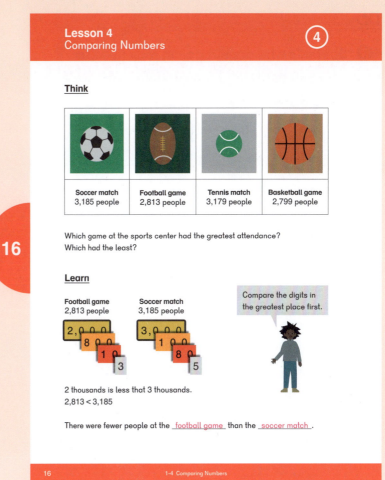

| Soccer match 3,185 people | Football game 2,813 people | Tennis match 3,179 people | Basketball game 2,799 people |

Which game at the sports center had the greatest attendance? Which had the least?

Learn

Football game 2,813 people   Soccer match 3,185 people

Compare the digits in the greatest place first.

2 thousands is less that 3 thousands.
2,813 < 3,185

There were fewer people at the _football game_ than the _soccer match_.

16   1-4 Comparing Numbers

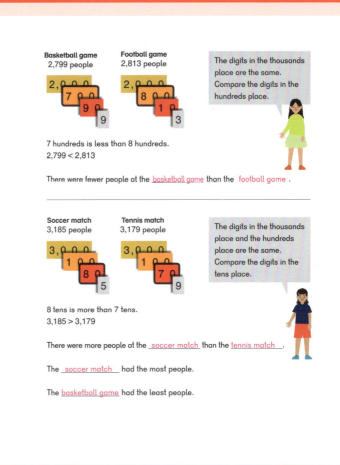

Basketball game 2,799 people   Football game 2,813 people

The digits in the thousands place are the same. Compare the digits in the hundreds place.

7 hundreds is less than 8 hundreds.
2,799 < 2,813

There were fewer people at the _basketball game_ than the _football game_.

Soccer match 3,185 people   Tennis match 3,179 people

The digits in the thousands place and the hundreds place are the same. Compare the digits in the tens place.

8 tens is more than 7 tens.
3,185 > 3,179

There were more people at the _soccer match_ than the _tennis match_.

The _soccer match_ had the most people.

The _basketball game_ had the least people.

1-4 Comparing Numbers   17

Do

❶—❸ Students who are struggling may need to make the numbers with Place-value Cards (BLM) to compare the place values.

❹ Ask students if there is a way to compare the numbers without adding the numbers first.

Students can consider the place values from greatest to least. In **❹**(a), for example, both numbers begin with 7,000. There are 8 hundreds in the first number and the digit in the hundreds place in the second number is 9, which has a value of 9 hundreds.

$7,000 + 800 < 7,900$

❺ Note that Alex is telling us that 0,873 is not a four-digit number. A number cannot begin with zero, unless the number is 0.

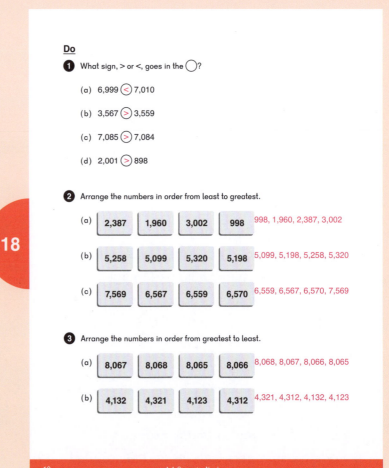

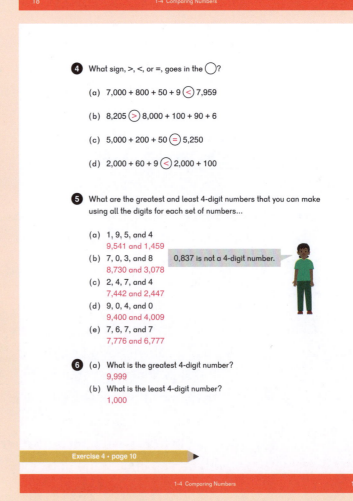

Activity

▲ Greatest or Least?

Materials: Number Cards (BLM) 0 to 9 or a 10-sided die, whiteboards, dry erase markers

Whole class play:

* Players draw a four-digit place-value chart plus one extra box labeled "trash" as shown below.

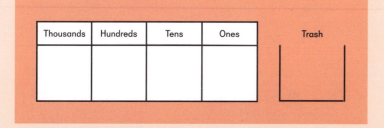

* The teacher draws a Number Card (BLM). Each player decides in which place to write the number before the next card is chosen.
* Once the digit has been written, it must stay in that place.
* Players may throw one number in the trash.
* When 5 cards have been drawn, the players with the greatest number are the winners.

Play in groups up to 4 players:

* Players take turns drawing Number Cards (BLM) and writing the number in one of their places.
* After all players have drawn 5 Number Cards (BLM), they compare their numbers. The player with the greatest four-digit number wins.

Modify to play for the least four-digit number.

Exercise 4 • page 10

Lesson 5 The Number Line

Objective

- Locate numbers on a number line.

Lesson Materials

- String or yarn, 12 feet
- Note cards, folded in half with multiples of 10 to 100: 0, 10, 20, 30, ... 90, 100
- Note cards, folded in half with multiples of 100 to 1,000: 0, 100, 200, ... 900, 1,000
- Number Line (BLM)
- Dry erase sleeves

Think

Create an interactive line of numbers. Have two students hold the two ends of a string, or affix it to the board or wall in a classroom. Ask students to place the prepared 0 and 100 note cards on the string.

Ask a student to put the number 10 card on the string where it belongs, moving the 0 and 100 cards if necessary.

Continue handing out cards until all multiples of ten cards have been placed on the string.

Discuss the difference in value between the number cards on the string.

Remove the cards and repeat the activity with the multiples of 100 cards.

The 0 and 1,000 should be in the same place that the 0 and 100 were before. Ask students, "What's the difference in value between the numbers now?"

Have students solve the **Think** problem and share their methods.

Ask students how the number lines on textbook page 20 are similar and different from the activity with the string.

Finally, discuss what is different about the red number line. Students should note that it starts not with 0, but with the number 1,000.

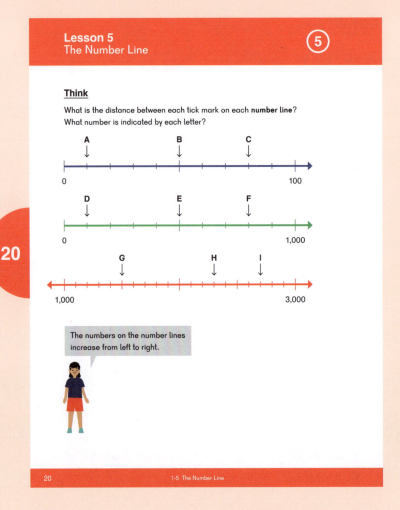

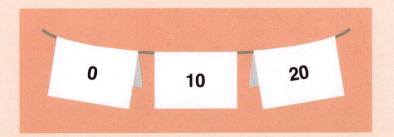

Learn

Distances between tick marks on a number line are referred to as increments. Have students count the increments as they hop along the number line with their fingers. Remind them not to count the tick marks they are on, but to count the hops, like in a board game.

Have students discuss the numbers that are halfway between tick marks.

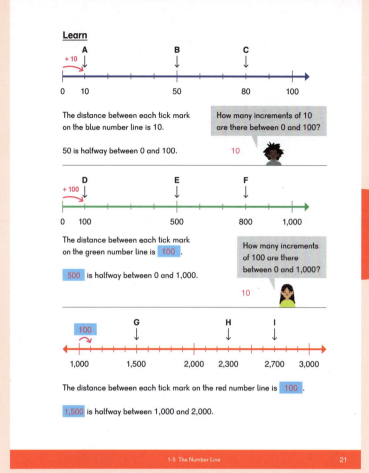

Do

Students can use a Number Line (BLM) in a dry erase sleeve and label each tick mark.

Students should note that number lines always go from least to greatest as they go from left to right. A number to the **right** of another on a number line is always the **greater** number. A number to the **left** of another number is always the **lesser** number.

Locating numbers on a number line is another way of comparing numbers.

Number lines are a student's first introduction to the x-axis in a Cartesian plane.

In ❹ and ❺, students will look at numbers that are near, but not halfway between tick marks.

❹ — ❻ Discuss Emma's comment about point M on the number line. Have students share where they think the letters belong on these number lines and why. In ❹, M and N point to a spot near a tick mark.

❻ Use Number Line (BLM) or have students draw a number line on a whiteboard to show where each number belongs.

Exercise 5 • page 13

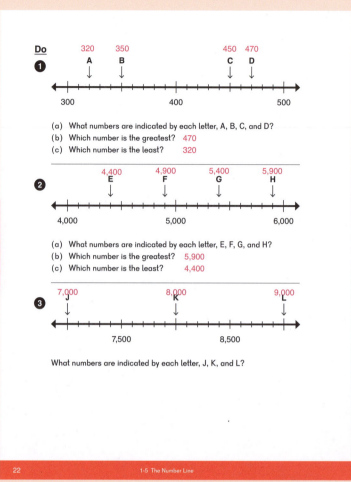

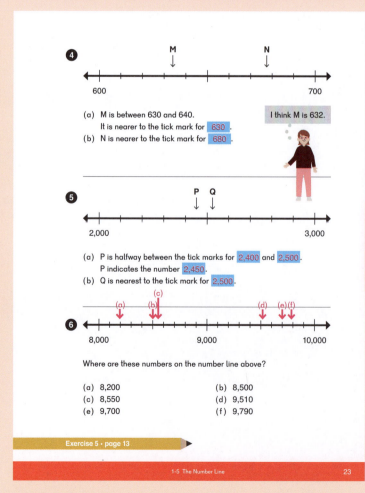

Objective

• Practice concepts from the chapter.

After students complete the **Practice** in the textbook, have them continue working with place value by playing games from the chapter.

A solid understanding of four-digit numbers is necessary before students move on to rounding numbers.

❹ Extend by having students research the heights of other buildings in the world.

❺—❼ Extend by having students write similar problems and letting their classmates find the solutions.

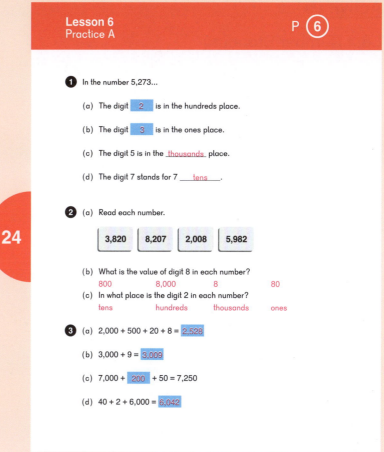

Lesson 6
Practice A   P ⑥

❶ In the number 5,273...

   (a) The digit [2] is in the hundreds place.

   (b) The digit [3] is in the ones place.

   (c) The digit 5 is in the _thousands_ place.

   (d) The digit 7 stands for 7 ___tens___ .

❷ (a) Read each number.

   | 3,820 | 8,207 | 2,008 | 5,982 |

   (b) What is the value of digit 8 in each number?
     800   8,000   8   80

   (c) In what place is the digit 2 in each number?
     tens   hundreds   thousands   ones

❸ (a) 2,000 + 500 + 20 + 8 = [2,528]

   (b) 3,000 + 9 = [3,009]

   (c) 7,000 + [200] + 50 = 7,250

   (d) 40 + 2 + 6,000 = [6,042]

24   1-6 Practice A

24

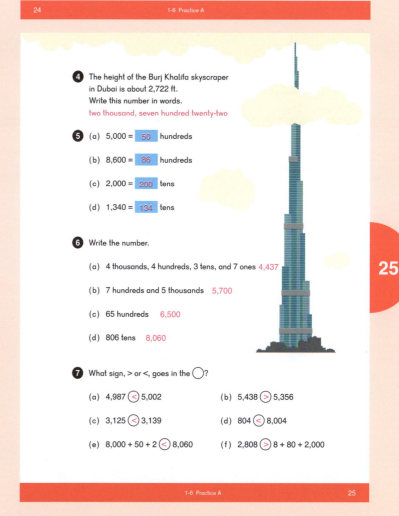

❹ The height of the Burj Khalifa skyscraper in Dubai is about 2,722 ft. Write this number in words.
two thousand, seven hundred twenty-two

❺ (a) 5,000 = [50] hundreds

   (b) 8,600 = [86] hundreds

   (c) 2,000 = [200] tens

   (d) 1,340 = [134] tens

❻ Write the number.

   (a) 4 thousands, 4 hundreds, 3 tens, and 7 ones 4,437

   (b) 7 hundreds and 5 thousands 5,700

   (c) 65 hundreds 6,500

   (d) 806 tens 8,060

❼ What sign, > or <, goes in the ◯?

   (a) 4,987 ⊙< 5,002   (b) 5,438 ⊙> 5,356

   (c) 3,125 ⊙< 3,139   (d) 804 ⊙< 8,004

   (e) 8,000 + 50 + 2 ⊙< 8,060   (f) 2,808 ⊙> 8 + 80 + 2,000

1-6 Practice A   25

25

Activities

▲ Last Number Standing

This is a whole class activity.

Have students write a four-digit number on paper or their whiteboard and then stand. Give students place-value amounts:

- If you have the digit 4 in the thousands place, sit down.
- If your number has 13 tens, sit down.
- If the value of your tens place is halfway between one and one hundred, sit down.

Students who sit down can share their numbers aloud. Have them predict the last number standing.

Call out place-value amounts until only one student is left standing.

▲ My Name Is...

Materials: Name tags with decomposed numbers written on them, recording sheet listing names of all students in the class

Give each player a name tag showing a four-digit number in decomposed form.

Version 1: Provide a recording sheet with students' names and have students record their classmates' number next to their names.

Task students to find out who has the greatest number and the least number.

Version 2: Ensure that there are 2 name tags that represent the same amount. Have students find their match.

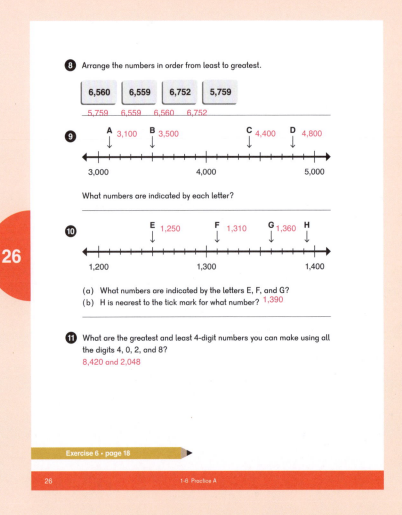

(8) Arrange the numbers in order from least to greatest.

| 6,560 | 6,559 | 6,752 | 5,759 |

5,759   6,559   6,560   6,752

(9)   A 3,100   B 3,500   C 4,400   D 4,800

3,000   4,000   5,000

What numbers are indicated by each letter?

(10)   E 1,250   F 1,310   G 1,360   H

1,200   1,300   1,400

(a) What numbers are indicated by the letters E, F, and G?
(b) H is nearest to the tick mark for what number? 1,390

(11) What are the greatest and least 4-digit numbers you can make using all the digits 4, 0, 2, and 8? 8,420 and 2,048

Exercise 6 • page 18

26   1-6 Practice A

Exercise 6 • page 18

Lesson 7 Number Patterns

Objective

- Find a number that is 10, 100, or 1,000 more or less than a four-digit number.

Lesson Materials

- Place-value discs

Think

Provide students with place-value discs and have them try to solve the **Think** problem independently.

Discuss student strategies for solving the problem.

Note that the emphasis is on paying attention to the place value of the discs that are being added or subtracted. This determines which digits change.

Examples:

- I made the numbers with discs, then I added a 1,000-disc, then I took off a 100-disc.
- I lined the numbers up like I was taught in grade 2.
- I knew 1,000 more than 6,000 was 7,000.

Students may find alternative strategies.

Learn

Have students show the number of stickers that Sofia had to start with, using place-value discs. Have them add one 1,000-disc to find the amount of stickers she has after her mother gives her 1,000 stickers.

Have students take one 100-disc away to find the amount of stickers that Sofia has after giving 100 stickers to a friend.

Do

Students who are struggling may need to use place-value discs. Most students should be working with just the pictures in the textbook.

2 Mei is pointing out that if we think of 19 hundreds + 1 hundred = 20 hundreds, we have learned that 20 hundreds = 2 thousands.

3 Alex and Emma remind us that 6,000 is the same as 60 hundreds or 600 tens.

Have students write the numbers on a whiteboard. This skill is important in developing number sense. Students who struggle should continue to work these problems with place-value discs.

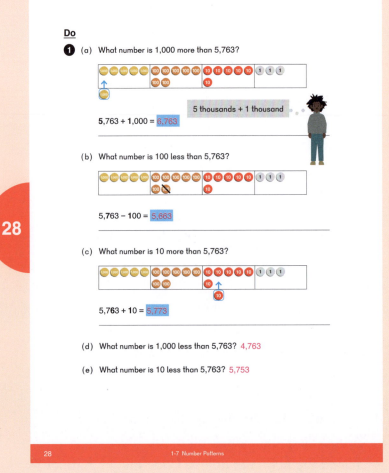

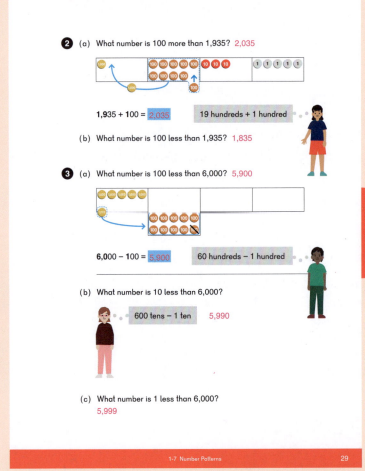

Activities

▲ Choral Counting

Use this whole class activity as a warm-up or review throughout the year. Using your thumb to point up or down, have students chorally count on and back by ones, tens, or hundreds.

Example: You say, "Let's count by tens starting at 2,488. First number?" Class responds, "2,488." Point thumb up. Class responds, "2,498." Point thumb up. Class responds, "2,508." Point down. Class responds, "2,498."

▲ What's My Rule?

Students take turns being the Ruler and the Solver.

The Ruler writes a four-digit number and a rule for generating a sequence of numbers.

The Solver writes the next five numbers that follow the rule. For example:

The ruler writes, "5,687" and, "+ 100."

The solver needs to find 5,797, 5,897, 5,997, 6,097, and 6,197.

★ Challenge students with a two-step rule. For example:

The ruler writes, "2,387" and "+ 100, − 10."

The solver needs to find 2,477, 2,567, 2,657, 2,747, and 2,837.

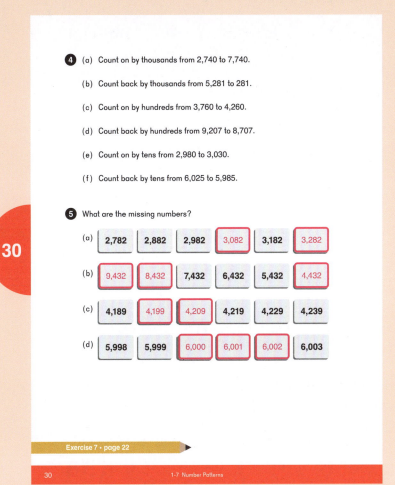

Exercise 7 • page 22

Objective

- Round a four-digit number to the nearest thousand.

Lesson Materials

- Number Line (BLM)
- Dry erase sleeves

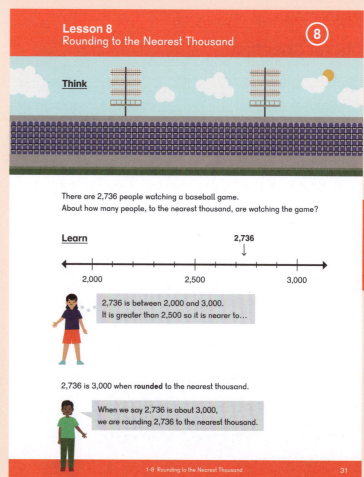

Think

Provide each student with a Number Line (BLM) in a dry erase sleeve. Give students a four-digit number. Ask them to show both the nearest thousand numbers that are less than and greater than the given number.

Repeat until students can accurately find the numbers.

Pose the **Think** problem. Have students recall the lesson on number lines (Lesson 5 on page 18 of this Teacher's Guide). Ask students, "What does it mean when we say the number 2,736 is nearer to 3,000 than 2,000?"

Learn

Discuss how students would find the thousands numbers on their number line. 2,000 is the nearest thousand less than 2,736 and 3,000 is the nearest thousand greater than 2,736. Have them use a blank Number Line (BLM) in a dry erase sleeve.

If students struggle with the number line activity as described, have them practice by creating a student number line.

If a student is working with the number 2,736, he will start by identifying which of his classmates are holding the closest multiples of 1,000 (2,000 and 3,000). He will then stand between them, as the friends are showing in this illustration, to show that 2,736 is between 2,000 and 3,000.

Do

❶ (b) Dion is reminding us of the convention that when we are halfway between two thousands, or equally far from both benchmarks, we round to the greater thousand.

❸ Have students use a blank Number Line (BLM) in a dry erase sleeve.

Activity

▲ Shut the Box — Rounding

Materials: Shut the Box — Thousands Game Board (BLM), several sets of Number Cards (BLM) 0 to 9 or 4 10-sided dice, 10 counters

Players take turns either drawing 4 Number Cards (BLM) or rolling 4 dice. On each turn:

- Players form a four-digit number with the Number Cards (BLM) drawn.
- They round that number to the nearest thousand.
- They cover that number with a counter on their Shut the Box — Thousands Game Board (BLM).

If the number is already covered, a player's turn is over. The winner is the first player to cover all of the numbers on his Shut the Box — Thousands Game Board (BLM).

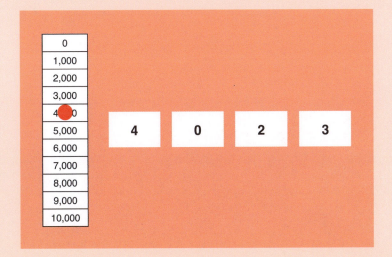

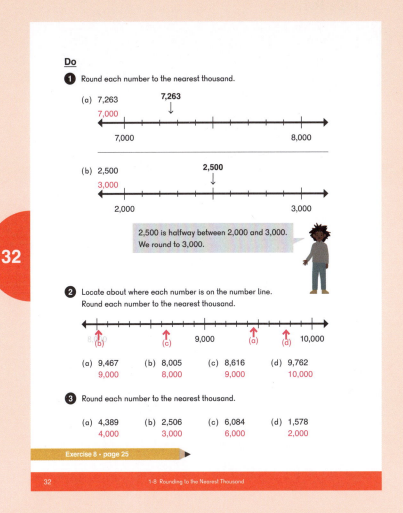

Exercise 8 • page 25

Lesson 9 Rounding to the Nearest Hundred

Objective

- Round a four-digit number to the nearest hundred.

Lesson Materials

- Number Lines (BLM)
- Dry erase sleeves

Think

Provide each student with a Number Line (BLM) in a dry erase sleeve and a four-digit number. Have them show the number that is a multiple of 100 that comes before and after the four-digit number given.

Once they have done that, they can round to the nearest hundred.

Pose the **Think** problems. Have students show their number on the Number Line (BLM) and explain their solutions to the problems.

Learn

Discuss how students would find the two hundreds that 2,736 is between.

2,700 is the nearest hundred less than 2,736 and 2,800 is the nearest hundred greater than 2,736.

Sofia reminds us that when a number is halfway between two benchmark numbers, it should be rounded to the greater number. 2,750 rounds to 2,800.

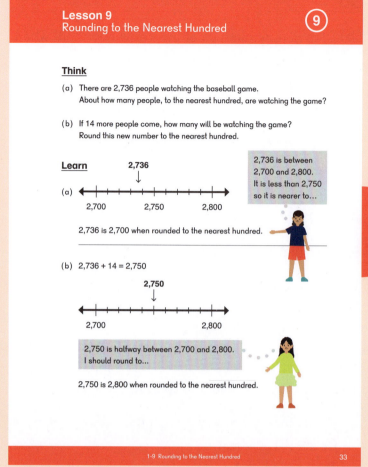

Do

1 (b) Alex is reminding students that 2,000 = 20 hundreds, so 1,997 is between 19 hundreds and 20 hundreds.

2 (a) Have students use a Number Line (BLM) in a dry erase sleeve to show where each number should be located on the number line.

Activity

▲ Three in a Row — Rounding

Materials: Three in a Row — Rounding to Hundreds Game Board (BLM) in a dry erase sleeve, several sets of Number Cards (BLM) 0 to 9 or 4 10-sided dice, dry erase markers

Players take turns either drawing 4 Number Cards (BLM) or rolling 4 dice. On each turn:

- Players form a four-digit number with the Number Cards (BLM) drawn.
- They round that number to the nearest hundred.
- They mark out that number on their Three in a Row — Rounding to Hundreds Game Board (BLM).

If the number is already covered, the player's turn is over. The winner is the first player to cover three of her numbers in a row horizontally, vertically, or diagonally on her Three in a Row — Rounding to Hundreds Game Board (BLM).

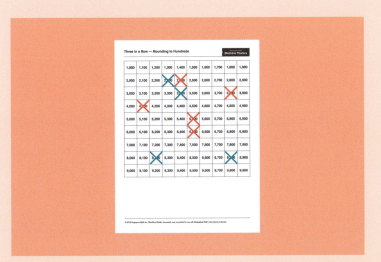

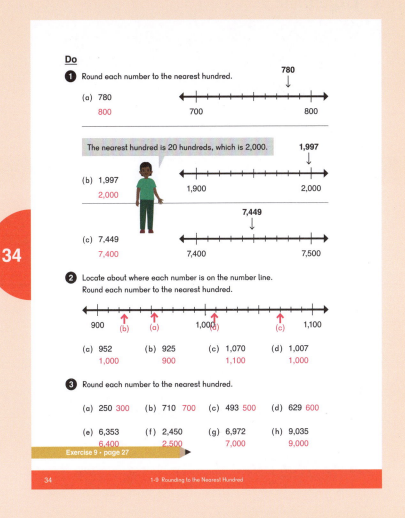

Exercise 9 • page 27

Lesson 10 Rounding to the Nearest Ten

Objective

- Round a four-digit number to the nearest ten.

Lesson Materials

- Number Line (BLM)
- Dry erase sleeves

Think

Provide each student with a Number Line (BLM) in a dry erase sleeve. Give them two-digit and three-digit numbers prior to having them round four-digit numbers to the nearest ten. Have students show both the nearest tens numbers that are less than and greater than the given number.

Once they have done that, they can round four-digit numbers to the nearest ten.

Pose the **Think** problems. Have students show their numbers on a Number Line (BLM) and explain their solutions to the problems.

Learn

Discuss how students would find the two tens that 2,376 is between.

2,730 is the nearest ten less than 2,736 and 2,740 is the nearest ten greater than 2,736.

1,004 is nearer to the benchmark number of 1,000 than it is to 1,010.

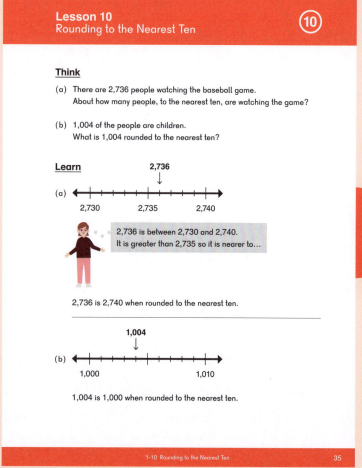

Do

❶ (a–b) and ❸ (a–h) provide problems with rounding to the nearest ten with two-digit and three-digit numbers. Students who are struggling with rounding to the nearest ten should be given more of these problems to practice before rounding four-digit numbers.

❷ (a) Have students use Number Lines (BLM) in dry erase sleeves to show where each number should be located on the number line.

Activity

▲ **Greatest or Least — Rounding**

Materials: Several sets of Number Cards (BLM) 0 to 9 or a 10-sided die, whiteboards, dry erase markers

Modify the rules from Lesson 4 on page 17 of this Teacher's Guide.

When players have a four-digit number, have them round it to the nearest ten, hundred, or thousand.

Exercise 10 • page 30

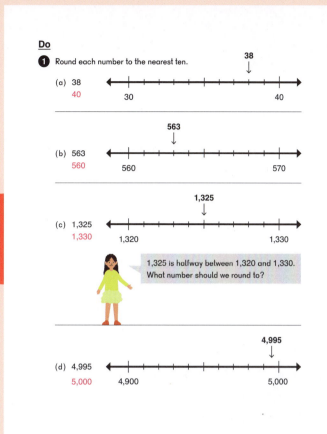

Do

❶ Round each number to the nearest ten.

(a) 38
40

(b) 563
560

(c) 1,325
1,330

1,325 is halfway between 1,320 and 1,330. What number should we round to?

(d) 4,995
5,000

❷ Locate about where each number is on the number line. Round each number to the nearest ten.

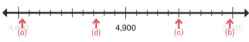

(a) 4,805 4,810   (b) 4,996 5,000

(c) 4,950 4,950   (d) 4,872 4,870

❸ Round each number to the nearest ten.

(a) 48   (b) 75   (c) 96
    50   80   100
(d) 23   (e) 231   (f) 897
    20   230   900
(g) 495   (h) 903   (i) 4,166
    500   900   4,170
(j) 1,585   (k) 7,097   (l) 3,996
    1,590   7,100   4,000

❹ Eli the elephant weighs 6,527 pounds. Round his weight to the nearest thousand, the nearest hundred, and the nearest ten.
nearest thousand: 7,000
nearest hundred: 6,500
nearest ten: 6,530

Exercise 10 • page 30

Lesson 11 Practice B

Objective

- Practice skills and concepts from the chapter.

After students complete the **Practice** in the textbook, have them continue working with place value and rounding by playing games from the chapter.

Activity

- ## Snowball Review

Materials: Blank paper

Write large four-digit numbers on sheets (or half sheets) of paper, enough for each student in the class to have 2 or 3. Underline either the thousands, hundreds, or tens place in each number to indicate to which place students will round the number.

Show students examples and explain that the underlined number is the place to which they will round.

Have each student crumple up their papers and have a classroom snowball fight for one minute. (No running, safety first!)

At the end of the fight, ask kids to grab a snowball and return to their seats. One at a time, have students unwrap the paper, show the class, and then round to the correct value.

When the class has each rounded one snowball number, have another "fight" and repeat with remaining snowballs on the floor.

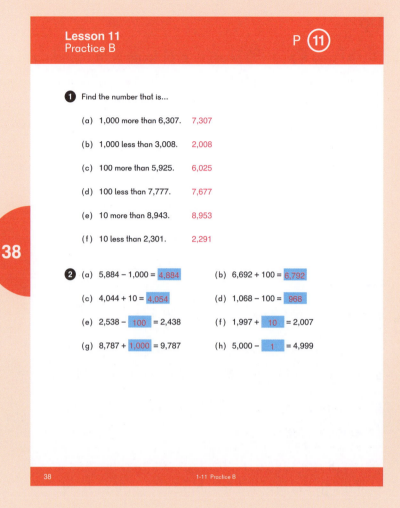

Brain Works

★ What Number Am I? Riddles

What number am I?

- I am a four-digit number.
- I am less than 6,000 but greater than 5,000.
- My hundreds digit is less than 7 but greater than 5.
- My tens digit is an odd number less than 5 but greater than 1.
- My ones digit is in the 3 times table and is greater than 4 but less than 8.

What number am I?

- I am a four-digit number.
- My thousands digit is the number of days in a week.
- My tens digit is the fingers on your hand.
- If you double my ones digit you get 12.
- My hundreds digit is $1 + 2 + 3 + 6 - 7$.

Have students create their own riddles and share.

Answers

5,636

7,556

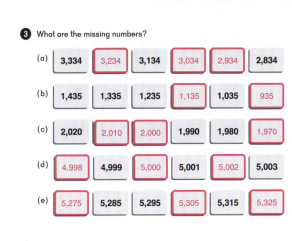

3. What are the missing numbers?

(a) 3,334 | 3,234 | 3,134 | 3,034 | 2,934 | 2,834

(b) 1,435 | 1,335 | 1,235 | 1,135 | 1,035 | 935

(c) 2,020 | 2,010 | 2,000 | 1,990 | 1,980 | 1,970

(d) 4,998 | 4,999 | 5,000 | 5,001 | 5,002 | 5,003

(e) 5,275 | 5,285 | 5,295 | 5,305 | 5,315 | 5,325

4. (a) Count on by thousands from 2,910 to 7,910.

(b) Count back by thousands from 5,371 to 371.

(c) Count on by hundreds from 4,780 to 5,280.

(d) Count back by hundreds from 8,105 to 7,605.

39

1-11 Practice B   39

5. Locate 7,365 on each number line.
Then round 7,365 to the nearest...

(a) Thousand
7,000   7,000   8,000

(b) Hundred
7,400   7,300   7,400

(c) Ten
7,370   7,360   7,370

40

6. Round each number to the nearest thousand.

(a) 4,209   (b) 2,070   (c) 3,505   (d) 900
4,000   2,000   4,000   1,000

7. Round each number to the nearest hundred.

(a) 670   (b) 3,250   (c) 9,084   (d) 5,555
700   3,300   9,100   5,600

8. Round each number to the nearest ten.

(a) 163   (b) 3,287   (c) 8,005   (d) 4,996
160   3,290   8,010   5,000

40   1-11 Practice B

Chapter 1 Numbers to 10,000

Exercise 1

Basics

1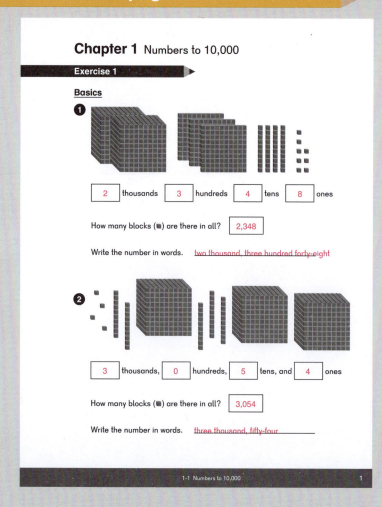

| 2 | thousands | 3 | hundreds | 4 | tens | 8 | ones |

How many blocks (▪) are there in all? 2,348

Write the number in words. *two thousand, three hundred forty-eight*

2

| 3 | thousands, | 0 | hundreds, | 5 | tens, and | 4 | ones |

How many blocks (▪) are there in all? 3,054

Write the number in words. *three thousand, fifty-four*

---

3 (a) 6,000 + 300 + 20 + 4 = [6,324]

(b) 7,000 + 500 + 20 = [7,520]

Practice

4 Write the number and the number word.

(a)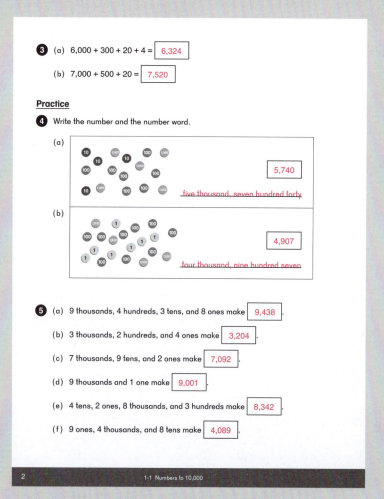

[5,740]

five thousand, seven hundred forty

(b)

[4,907]

four thousand, nine hundred seven

5 (a) 9 thousands, 4 hundreds, 3 tens, and 8 ones make [9,438].

(b) 3 thousands, 2 hundreds, and 4 ones make [3,204].

(c) 7 thousands, 9 tens, and 2 ones make [7,092].

(d) 9 thousands and 1 one make [9,001].

(e) 4 tens, 2 ones, 8 thousands, and 3 hundreds make [8,342].

(f) 9 ones, 4 thousands, and 8 tens make [4,089].

---

6 (a) 8,000 + 900 + 7 = [8,907]

(b) 2,000 + 20 = [2,020]

(c) 20 + 5,000 + 300 + 7 = [5,327]

(d) 400 + 3,000 + 30 + 4 = [3,434]

7 Write the number.

eight thousand, four hundred twenty-two	8,422
nine thousand, thirty-seven	9,037
five thousand, three hundred seven	5,307
two thousand, six	2,006

8 Count on by ones and write the missing numbers.

(a) | 997 | 998 | 999 | 1,000 | 1,001 | 1,002 |

(b) | 5,896 | 5,897 | 5,898 | 5,899 | 5,900 | 5,901 |

Challenge

9 Count on by threes and write the missing numbers.

| 4,295 | 4,298 | 4,301 | 4,304 | 4,307 | 4,310 |

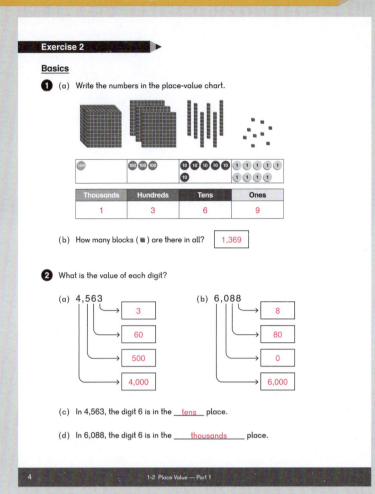

Exercise 2

Basics

1 (a) Write the numbers in the place-value chart.

Thousands	Hundreds	Tens	Ones
1	3	6	9

(b) How many blocks ( ) are there in all? **1,369**

2 What is the value of each digit?

(a) 4,563

→ 3
→ 60
→ 500
→ 4,000

(b) 6,088

→ 8
→ 80
→ 0
→ 6,000

(c) In 4,563, the digit 6 is in the **tens** place.

(d) In 6,088, the digit 6 is in the **thousands** place.

Practice

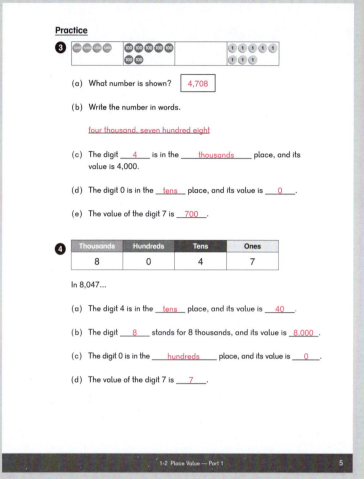

3

(a) What number is shown? **4,708**

(b) Write the number in words.

four thousand, seven hundred eight

(c) The digit **4** is in the **thousands** place, and its value is 4,000.

(d) The digit 0 is in the **tens** place, and its value is **0**.

(e) The value of the digit 7 is **700**.

4

Thousands	Hundreds	Tens	Ones
8	0	4	7

In 8,047...

(a) The digit 4 is in the **tens** place, and its value is **40**.

(b) The digit **8** stands for 8 thousands, and its value is **8,000**.

(c) The digit 0 is in the **hundreds** place, and its value is **0**.

(d) The value of the digit 7 is **7**.

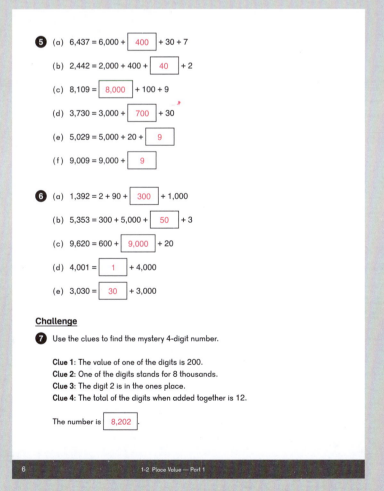

5 (a) 6,437 = 6,000 + **400** + 30 + 7

(b) 2,442 = 2,000 + 400 + **40** + 2

(c) 8,109 = **8,000** + 100 + 9

(d) 3,730 = 3,000 + **700** + 30

(e) 5,029 = 5,000 + 20 + **9**

(f) 9,009 = 9,000 + **9**

6 (a) 1,392 = 2 + 90 + **300** + 1,000

(b) 5,353 = 300 + 5,000 + **50** + 3

(c) 9,620 = 600 + **9,000** + 20

(d) 4,001 = **1** + 4,000

(e) 3,030 = **30** + 3,000

Challenge

7 Use the clues to find the mystery 4-digit number.

Clue 1: The value of one of the digits is 200.
Clue 2: One of the digits stands for 8 thousands.
Clue 3: The digit 2 is in the ones place.
Clue 4: The total of the digits when added together is 12.

The number is **8,202** .

Teacher's Guide 3A Chapter 1   35

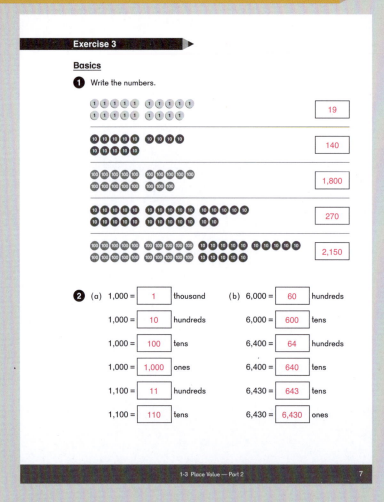

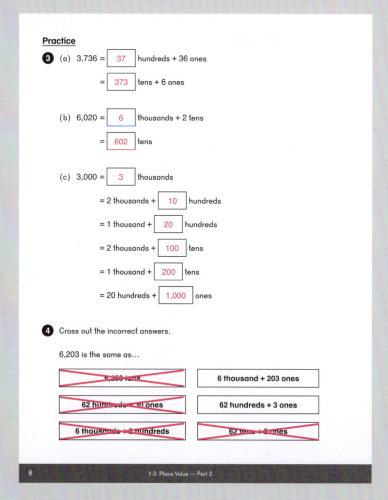

Exercise 3

Basics

1. Write the numbers.

(ones)	19
(tens)	140
(hundreds)	1,800
(tens)	270
(hundreds/tens)	2,150

2. (a) 1,000 = [1] thousand
   1,000 = [10] hundreds
   1,000 = [100] tens
   1,000 = [1,000] ones
   1,100 = [11] hundreds
   1,100 = [110] tens

   (b) 6,000 = [60] hundreds
   6,000 = [600] tens
   6,400 = [64] hundreds
   6,400 = [640] tens
   6,430 = [643] tens
   6,430 = [6,430] ones

1-3 Place Value — Part 2   7

Practice

3. (a) 3,736 = [37] hundreds + 36 ones
   = [373] tens + 6 ones

   (b) 6,020 = [6] thousands + 2 tens
   = [602] tens

   (c) 3,000 = [3] thousands
   = 2 thousands + [10] hundreds
   = 1 thousand + [20] hundreds
   = 2 thousands + [100] tens
   = 1 thousand + [200] tens
   = 20 hundreds + [1,000] ones

4. Cross out the incorrect answers.

   6,203 is the same as…

~~6,203 tens~~	6 thousand + 203 ones
~~62 hundreds + 3 ones~~ (partly crossed)	62 hundreds + 3 ones
6 thousands + 2 hundreds	~~62 tens + 3 ones~~

8   1-3 Place Value — Part 2

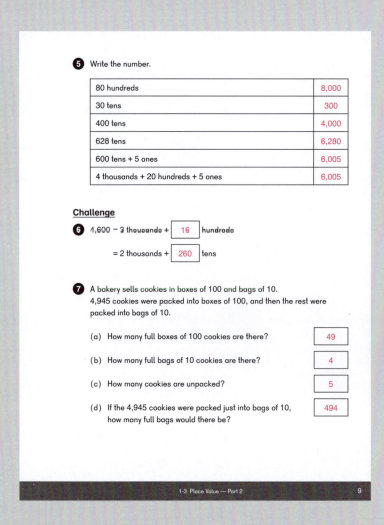

5. Write the number.

80 hundreds	8,000
30 tens	300
400 tens	4,000
628 tens	6,280
600 tens + 5 ones	6,005
4 thousands + 20 hundreds + 5 ones	6,005

Challenge

6. 4,600 = 2 thousands + [16] hundreds
   = 2 thousands + [260] tens

7. A bakery sells cookies in boxes of 100 and bags of 10.
   4,945 cookies were packed into boxes of 100, and then the rest were packed into bags of 10.

   (a) How many full boxes of 100 cookies are there? [49]

   (b) How many full bags of 10 cookies are there? [4]

   (c) How many cookies are unpacked? [5]

   (d) If the 4,945 cookies were packed just into bags of 10, how many full bags would there be? [494]

1-3 Place Value — Part 2   9

Exercise 4

Basics

1 Which number is greater?
Circle it.

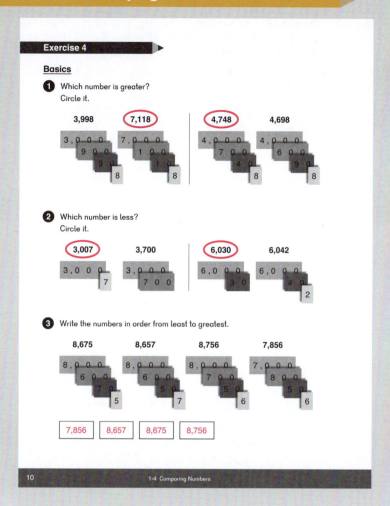

3,998   **(7,118)** | **(4,748)**   4,698

2 Which number is less?
Circle it.

(3,007)   3,700 | **(6,030)**   6,042

3 Write the numbers in order from least to greatest.

8,675   8,657   8,756   7,856

| 7,856 | 8,657 | 8,675 | 8,756 |

Practice

4 Write > or < in the ○.

(a) 8,262 **(>)** 2,558   (b) 9,532 **(>)** 9,352

(c) 6,365 **(<)** 6,390   (d) 5,556 **(<)** 5,565

5 Write the numbers in order from greatest to least.

5,900   9,050 | 9,050 | 5,900 | 5,009 | 950 |
      950   5,009

6 The table shows the heights of some mountains in meters.

Mountain	Meters
Pichu Pichu (Peru)	5,664
Little Si (U.S.A.)	480
Mount Tyree (Antarctica)	4,852
Mount Everest (Nepal)	8,848
Mont Blanc (France)	4,810
Castle Peak (U.S.A.)	4,348

List the mountains in order from the tallest to the shortest.

Mount Everest, Pichu Pichu, Mount Tyree, Mont Blanc,
Castle Peak, Little Si

---

7 Write the greatest and least 4-digit number you can make using all
the digits.

Digits	Greatest	Least
6, 2, 1, 5	6,521	1,256
4, 0, 6, 8	8,640	4,068

8 Use all of the digits 6, 8, 5, 2 to make…

(a) The greatest number between 2,700 and 5,700. | 5,682 |

(b) The least number between 2,700 and 5,700. | 2,856 |

9 Write >, <, or = in the ○.

(a) 1,400 + 4 **(<)** 1,000 + 40 + 600

(b) 900 + 50 + 4,000 **(>)** 4,000 + 5 + 90

Challenge

10 Read the clues.
Then circle the correct number.

Clue 1: The digit 4 is in the hundreds place.
Clue 2: There are more than 40 hundreds.
Clue 3: There are less than 450 tens.

| 5,440 | 3,400 | **(4,445)** | 4,500 |

Teacher's Guide 3A Chapter 1

Exercise 5

Basics

❶ Write the value of the increment between each tick mark.
Then write the number indicated by each arrow.

(a) The increment between tick marks is ___10___.

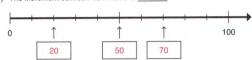

(b) The increment between tick marks is ___100___.

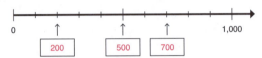

(c) The increment between tick marks is ___10___.

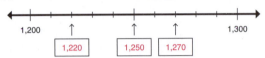

(d) The increment between tick marks is ___1___.

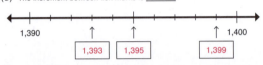

(e) The increment between tick marks is ___10___.

(f) The increment between tick marks is ___100___.

❷ For each number line, draw an arrow to the tick mark that is halfway
between the two labeled tick marks, and write the number.

(a)

___50___ is halfway between 0 and 100.

(b)

___550___ is halfway between 500 and 600.

(c)

___8,500___ is halfway between 8,000 and 9,000.

Practice

❸ Draw arrows to show the location of the numbers on each number line.
Label the arrows with the numbers.

(a) 25   11   19

(b) 850   980   920

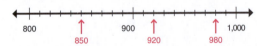

(c) 4,700   4,500   3,300

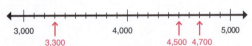

❹ For each number line, draw an arrow to the tick mark that is halfway
between the two labeled tick marks, and write the number.

(a)

___435___ is halfway between 430 and 440.

(b)

___2,345___ is halfway between 2,340 and 2,350.

❺ Draw arrows to show the approximate location of the numbers on
each number line.
Label the arrows with the numbers.

(a) 855   982   904

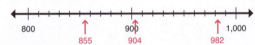

(b) 4,720   4,445   3,370

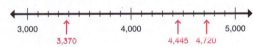

❻ Draw an arrow to show the location or approximate location of 1,825
on each of the following number lines.

(a)

(b)

(c)

Challenge

7 For each of the following, write the number indicated by each arrow. Also draw arrows and label them to indicate the location or approximate location of the given numbers.

(a) The increment between tick marks is ___2___.

142   151   159

_____

(b) The increment between tick marks is ___5___.

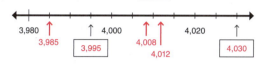

3,985   4,008   4,012

_____

(c) The increment between tick marks is ___200___.

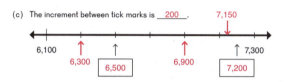

6,900   6,300   7,150

Exercise 6

Check

1 In the number 6,940...

(a) The digit [4] is in the tens place.

(b) The digit 9 is in the ___hundreds___ place.

(c) The digit 6 stands for 6 ___thousands___.

(d) The number is the same as [69] hundreds and [4] tens.

(e) Write the number in words.

six thousand, nine hundred forty

2 (a) 4,004 = [4] thousands + 4 ones

= [40] hundreds + 4 ones

= [400] tens + 4 ones

= [4,004] ones

(b) 8,904 = 8 thousands + [90] tens + 4 ones

= [89] hundreds + 4 ones

= [890] tens + 4 ones

= [8,904] ones

3 Draw arrows and label them to show the location or approximate location of the given numbers on each number line.

(a) 86   93   99

The number that is halfway between 80 and 90: ___85___

(b) 308   375   450

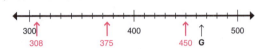

What number is indicated by G? ___465___

(c) 550   1,800   742

The number that is halfway between 1,000 and 2,000: ___1,500___

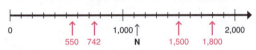

What number is indicated by the tick mark N is closest to? ___1,100___

4 (a) Write the numbers.

9,000 + 400 + 8	9,408
5 + 600 + 40 + 2,000	2,645
three thousand, eight hundred seventy-two	3,872
seven thousand, sixty-four	7,064
one thousand, one	1,001
3 thousand + 20 tens + 5 ones	3,205
350 tens	3,500
90 hundreds + 48 ones	9,048
700 tens + 7 tens	7,070

(b) Arrange the numbers above from greatest to least.

9,408 9,048 7,070 7,064 3,872 3,500 3,205 2,645 1,001

5 Write >, <, or = in the ◯.

(a) 400 tens + 20 tens ◯= 200 ones + 4 thousands

(b) 7 thousands + 23 tens ◯> 70 hundreds + 23 ones

(c) 34 hundreds + 20 tens ◯= 3,600

Challenge

6 A bank teller counted the money she had collected at the end of the day. She collected the following bills:

• 65 one-hundred dollar bills   6,500
• 23 ten-dollar bills   230
• 63 one-dollar bills   63

How much money did she collect?   [$6,793]

7 Use the clues and circle the correct number.

Clue 1: The difference between the digit in the hundreds place and the ones place is 6.
Clue 2: The digit 2 is in the tens place.
Clue 3: There are at most 7,000 ones.

[8,721]   (6,923)   [6,922]   [5,892]

8 Use the clues to find the mystery 4-digit number.

Clue 1: The digit in the tens place is twice the digit in the thousands place.
Clue 2: The digit in the hundreds place is 1 less than the digit in the thousands place.
Clue 3: The digit in the ones place is 2.
Clue 4: The sum of the digits is 13.

What is the number?   [3,262]

The digits in the thousands, hundreds and tens place have a sum of 11. Student can use logical trial and error to find those digits.

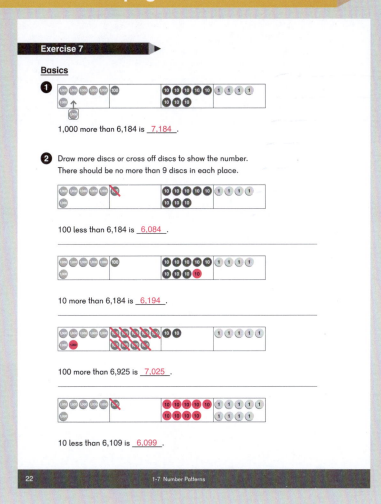

Exercise 7

Basics

1

1,000 more than 6,184 is __7,184__.

2 Draw more discs or cross off discs to show the number. There should be no more than 9 discs in each place.

100 less than 6,184 is __6,084__.

10 more than 6,184 is __6,194__.

100 more than 6,925 is __7,025__.

10 less than 6,109 is __6,099__.

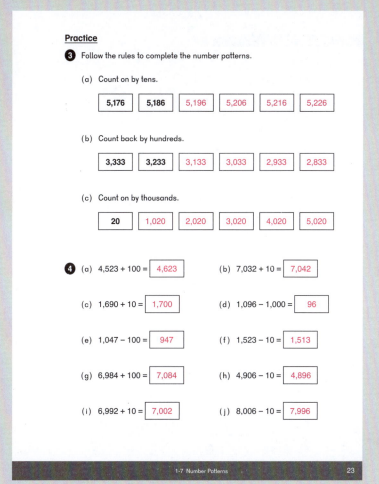

Practice

3 Follow the rules to complete the number patterns.

(a) Count on by tens.

| 5,176 | 5,186 | 5,196 | 5,206 | 5,216 | 5,226 |

(b) Count back by hundreds.

| 3,333 | 3,233 | 3,133 | 3,033 | 2,933 | 2,833 |

(c) Count on by thousands.

| 20 | 1,020 | 2,020 | 3,020 | 4,020 | 5,020 |

4 (a) $4,523 + 100 = \boxed{4,623}$ (b) $7,032 + 10 = \boxed{7,042}$

(c) $1,690 + 10 = \boxed{1,700}$ (d) $1,096 - 1,000 = \boxed{96}$

(e) $1,047 - 100 = \boxed{947}$ (f) $1,523 - 10 = \boxed{1,513}$

(g) $6,984 + 100 = \boxed{7,084}$ (h) $4,906 - 10 = \boxed{4,896}$

(i) $6,992 + 10 = \boxed{7,002}$ (j) $8,006 - 10 = \boxed{7,996}$

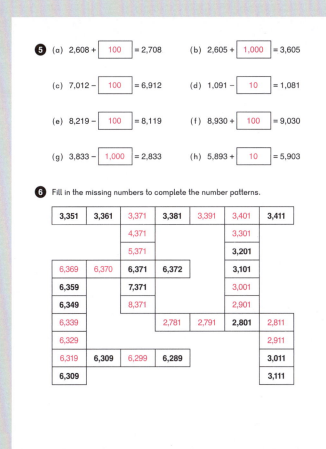

5 (a) $2,608 + \boxed{100} = 2,708$ (b) $2,605 + \boxed{1,000} = 3,605$

(c) $7,012 - \boxed{100} = 6,912$ (d) $1,091 - \boxed{10} = 1,081$

(e) $8,219 - \boxed{100} = 8,119$ (f) $8,930 + \boxed{100} = 9,030$

(g) $3,833 - \boxed{1,000} = 2,833$ (h) $5,893 + \boxed{10} = 5,903$

6 Fill in the missing numbers to complete the number patterns.

3,351	3,361	3,371	3,381	3,391	3,401	3,411
		4,371			3,301	
		5,371			3,201	
6,369	6,370	6,371	6,372		3,101	
6,359		7,371			3,001	
6,349		8,371			2,901	
6,339			2,781	2,791	2,801	2,811
6,329						2,911
6,319	6,309	6,299	6,289			3,011
6,309						3,111

Exercise 8

Basics

1 Fill in the blanks.

6,200   6,900   7,500

6,000   7,000   8,000

(a) 6,200 is 6,000 when rounded to the nearest ___thousand___.

(b) 6,900 is _7,000_ when rounded to the nearest thousand.

(c) 7,500 is halfway between 7,000 and _8,000_.

7,500 is _8,000_ when rounded to the nearest thousand.

2 Fill in the blanks.

3,005   3,530   4,168   4,761

3,000   4,000   5,000

(a) 3,005 is _3,000_ when rounded to the nearest thousand.

(b) 3,530 is _4,000_ when rounded to the nearest thousand.

(c) 4,168 is _4,000_ when rounded to the nearest thousand.

(d) 4,761 is _5,000_ when rounded to the nearest thousand.

(e) Look at the digit in the __hundreds__ place when rounding to the nearest thousand.

Practice

3 Draw arrows to show the location or approximate location of each number on the number line.
Then round each number to the nearest thousand.

7,000   8,000   9,000   10,000

7,500   8,350   8,980   9,672

(a) 7,500 | 8,000 | (b) 8,350 | 8,000 |

(c) 8,980 | 9,000 | (d) 9,672 | 10,000 |

4 The largest meteorite crater in the U.S. is 4,150 feet across. Round this number to the nearest thousand. | 4,000 |

5 Round each number to the nearest thousand.

(a) 1,920 | 2,000 | (b) 6,500 | 7,000 |

(c) 2,499 | 2,000 | (d) 8,052 | 8,000 |

Challenge

6 What is the greatest whole number that is 5,000 when rounded to the nearest thousand? | 5,499 |

7 What is the least whole number that is 5,000 when rounded to the nearest thousand? | 4,500 |

Exercise 9

Basics

1 Fill in the blanks.

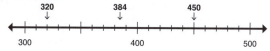

(a) 320 is 300 when rounded to the nearest ___hundred___.

(b) 384 is __400__ when rounded to the nearest hundred.

(c) 450 is halfway between 400 and __500__.

450 is __500__ when rounded to the nearest hundred.

2 Fill in the blanks.

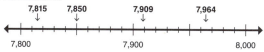

(a) 7,815 is __7,800__ when rounded to the nearest hundred.

(b) 7,850 is __7,900__ when rounded to the nearest hundred.

(c) 7,909 is __7,900__ when rounded to the nearest hundred.

(d) 7,964 is __8,000__ when rounded to the nearest hundred.

(e) Look at the digit in the ___tens___ place when rounding to the nearest hundred.

Practice

3 Round each number to the nearest hundred.

(a) 739 — 700   (b) 4,250 — 4,300

(c) 9,225 — 9,200   (d) 89 — 100

(e) 7,956 — 8,000   (f) 9,999 — 10,000

4 Round 6,565 to…

(a) The nearest thousand. — 7,000

(b) The nearest hundred. — 6,600

5 The table shows the heights of some mountains.
Round each number to the nearest thousand and hundred.

Mountain	Meters	Thousand	Hundred
Mount Everest	8,848	9,000	8,800
Cathedral Peak	3,326	3,000	3,300
Mount Stuart	2,869	3,000	2,900
Mount St. Helens	2,550	3,000	2,600
Sunset Peak	869	1,000	900

6 (a) Write the number that is halfway between 2,800 and 2,900, and the number that is halfway between 2,900 and 3,000.

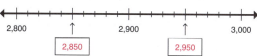

2,850   2,950

(b) What is the least whole number that is 2,900 when rounded to the nearest hundred?   2,850

(c) What is the greatest whole number that is 2,900 when rounded to the nearest hundred?   2,949

Challenge

7 A number between 280 and 380, when rounded to the nearest hundred, is 45 less than the original number. What number is the original number?   345

The number would have to round down to be less.
245 is not within the range, so it has to be 345.

Teacher's Guide 3A Chapter 1   43

Exercise 10

Basics

1 Fill in the blanks.

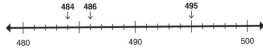

(a) 484 is 480 when rounded to the nearest __ten__.

(b) 486 is __490__ when rounded to the nearest ten.

(c) 495 is halfway between 490 and __500__.

495 is __500__ when rounded to the nearest ten.

2 Fill in the blanks.

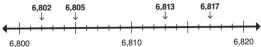

(a) 6,802 is __6,800__ when rounded to the nearest ten.

(b) 6,805 is __6,810__ when rounded to the nearest ten.

(c) 6,813 is __6,810__ when rounded to the nearest ten.

(d) 6,817 is __6,820__ when rounded to the nearest ten.

(e) Look at the digit in the ____ones____ place when rounding to the nearest ten.

Practice

3 Round each number to the nearest ten.

(a) 89 — 90

(b) 32 — 30

(c) 739 — 740

(d) 255 — 260

(e) 6,024 — 6,020

(f) 7,655 — 7,660

(g) 1,409 — 1,410

(h) 4,110 — 4,110

(i) 3,004 — 3,000

(j) 9,999 — 10,000

4 List the whole numbers that are 80 when rounded to the nearest ten.

75, 76, 77, 78, 79, 80, 81, 82, 83, 84

5 List the whole numbers that are 9,420 when rounded to the nearest ten.

9,415, 9,416, 9,417, 9,418, 9,419, 9,420

9,421, 9,422, 9,423, 9,424

6 A newborn baby blue whale weighs 2,865 pounds. Round this number to…

(a) The nearest thousand.   3,000

(b) The nearest hundred.   2,900

(c) The nearest ten.   2,870

7 A number is halfway between 8,420 and 8,430. Round this number to…

(a) The nearest thousand.   8,000

(b) The nearest hundred.   8,400

(c) The nearest ten.   8,430

Challenge

8 List the whole numbers that are 550 when rounded to the nearest ten and 500 when rounded to the nearest hundred.

545, 546, 547, 548, 549

9 A 2-digit whole number, when rounded to the nearest hundred, is 24 more than the original number. What is the is the original number?

76   A 2-digit number is less than 100.
24 less than 100 is 76.

Exercise 11

Check

1 Decide whether a rounded number or an exact number is likely in each of the following.

Write a check mark in the table for your answer.   *Answers may vary.*

	Rounded	Exact
The number of children in a park.	✓	
The number of students on the bus after a field trip.		✓
The number of books owned by a library.		✓
The number of books in a house.	✓	
The number of bees in a beehive.	✓	
The number of flowers in a field.	✓	
The number of stitches in a knitting pattern.		✓
The number of points scored in a game.		✓
The number of beads in a jar.	✓	

2 Complete the number patterns.

(a) 3,852 | 3,952 | 4,052 | 4,152 | 4,252 | 4,352

(b) 8,403 | 8,393 | 8,383 | 8,373 | 8,363 | 8,353

3 Write + or − in the ◯ and write the missing numbers.

(a) 7,432 ⊕ 10 = 7,442

(b) 6,492 ⊖ 1,000 = 5,492

(c) 8,219 ⊕ 100 = 8,319

(d) 2,605 ⊕ 1,000 = 3,605

(e) 1,801 ⊖ 10 = 1,791

(f) 8,012 ⊖ 100 = 7,912

4 The table shows the weight of some newborn animals at the zoo in grams. Round each number to the nearest thousand, hundred, and ten.

Animal	Grams	Thousands	Hundreds	Tens
African elephant	9,942	10,000	9,900	9,940
Giraffe	5,850	6,000	5,900	5,850
Zebra	4,148	4,000	4,100	4,150
Mountain goat	3,419	3,000	3,400	3,420
Gorilla	2,505	3,000	2,500	2,510
Sea otter	1,985	2,000	2,000	1,990
Leopard	1,142	1,000	1,100	1,140
Brown bear	945	1,000	900	950

5 (a) Round 6,484 to the nearest thousand.   6,000

(b) Round 6,484 to the nearest ten.   6,480

Then, round that new number to the nearest hundred.   6,500

Then, round that new number to the nearest thousand.   7,000

(c) Compare the final rounded numbers in (a) and (b). What is the difference between them?

The number rounded in steps is 1,000 more.

6 Use the following numbers to create the number patterns based on the given rule.

3,125 | 5,919 | 6,127 | 5,809

2,125 | 4,125 | 6,119 | 6,123

5,799 | 6,019 | 6,121 | 6,219

6,125 | 5,789 | 5,819 | 5,125

(a) Subtract 10 to get the next number.

5,829 | 5,819 | 5,809 | 5,799 | 5,789

(b) Add 100 to get the next number.   Look for numbers where the tens and ones are the same.

5,819 | 5,919 | 6,019 | 6,119 | 6,219

(c) Subtract 1,000 to get the next number.   Look for numbers where only the thousands are different.

6,125 | 5,125 | 4,125 | 3,125 | 2,125

(d) Add 2 to get the next number.   Start by looking for numbers where the thousands are the same.

6,119 | 6,121 | 6,123 | 6,125 | 6,127

Challenge

7 Look carefully at each digit to determine the pattern. Complete the pattern.

(a) The numbers increase by 1,000 and by 100.

3,152 | 4,252 | 5,352 | 6,452 | 7,552 | 8,652

(b) 8,403 | 8,313 | 8,223 | 8,133 | 8,043 | 7,953

The numbers decrease by 100 and increase by 10.

8 A number rounded to the nearest ten, hundred, or thousand is 5,000.

(a) What is the least possible number it could be?   4,995

(b) What is the greatest possible number it could be?   5,004

9 A number is rounded first to the nearest ten, then that new number is rounded to the nearest hundred, then that new number, when rounded to the nearest thousand, is 5,000.

(a) What is the least possible number it could be?   4,445

(b) What is the greatest possible number it could be?   5,444

Teacher's Guide 3A Chapter 1

Notes

Suggested number of class periods: 12–13

	Lesson	Page	Resources		Objectives
	Chapter Opener	p. 53	TB:	p. 41	Write expressions and equations.
1	Mental Addition — Part 1	p. 54	TB: WB:	p. 42 p. 37	Learn and apply different strategies for adding two-digit numbers mentally.
2	Mental Addition — Part 2	p. 56	TB: WB:	p. 44 p. 41	Mentally add three-digit numbers that are multiples of ten.
3	Mental Subtraction — Part 1	p. 58	TB: WB:	p. 46 p. 43	Subtract two-digit numbers mentally.
4	Mental Subtraction — Part 2	p. 60	TB: WB:	p. 48 p. 47	Mentally subtract a three-digit multiple of 10 from another three-digit multiple of 10.
5	Making 100 and 1,000	p. 62	TB: WB:	p. 50 p. 49	Add or subtract a two-digit number close to a multiple of 10 or a three-digit number close to a multiple of 100 to or from a number with up to three digits.
6	Strategies for Numbers Close to Hundreds	p. 65	TB: WB:	p. 53 p. 51	Add or subtract 97, 98, and 99.
7	Practice A	p. 67	TB: WB:	p. 55 p. 53	Practice adding and subtracting mentally.
8	Sum and Difference	p. 68	TB: WB:	p. 56 p. 55	Understand the meaning of sum and difference. Represent sums and differences using a model.
9	Word Problems — Part 1	p. 71	TB: WB:	p. 60 p. 60	Understand part-whole situations in addition and subtraction. Represent part-whole situations using bar models to solve problems.
10	Word Problems — Part 2	p. 74	TB: WB:	p. 64 p. 62	Understand comparison situations in addition and subtraction. Represent comparison situations using bar models to solve problems.
11	2-Step Word Problems	p. 77	TB: WB:	p. 68 p. 64	Represent multi-step problem situations with a bar model. Solve multi-step problems with addition and subtraction.
12	Practice B	p. 80	TB: WB:	p. 72 p. 68	Practice skills and concepts from the chapter.
	Workbook Solutions	p. 81			

In **Dimensions Math® 2A**, students learned to:

- Add and subtract three-digit numbers using the vertical algorithm.

This chapter teaches strategies for addition and subtraction, which review and build on those covered in **Dimensions Math® 1B** and **2B**:

Add by making the next ten	$27 + 48 = 30 + 45 = 75$ 3 45	$27 + 48 = 25 + 50 = 75$ 25 2	$762 + 9 = 761 + 10 = 771$ 761 1
Add the ones first	$762 + 9 = 760 + 11 = 771$ 760 2		
Add the tens first, then add the ones	$56 + 23 = 56 + 20 + 3 = 79$ 20 3		$45 + 26 = 45 + 20 + 6 = 71$ 20 6
Add 97, 98, or 99 by making the next hundred	$99 + 52 = 100 + 51 = 151$ 1 51		
Add 97, 98, or 99 by adding 100 and subtracting 1, 2, or 3 (over-adding)	100 99 1 $99 + 52 = 100 + 52 - 1 = 151$		
Subtract from tens first	$36 - 9 = 26 + 1 = 27$ 26 10	$36 - 9 = 6 + 21 = 27$ 6 30	$762 - 9 = 752 + 1 = 753$ 752 10
Subtract using known facts	$36 - 9 = 20 + 7 = 27$ 20 16	$762 - 9 = 750 + 3 = 753$ 750 12	
Subtract by decomposing the minuend (or subtract twice)	$36 - 9 = 36 - 6 - 3 = 27$ 6 3		$762 - 9 = 760 - 7 = 753$ 2 7
Subtract 97, 98, or 99 by subtracting 100 and adding 1, 2, or 3	100 99 1 $480 - 99 = 480 - 100 + 1 = 381$		

Mental math refers to mental strategies that leverage number sense. It is a way to make difficult computation easier. It does not mean that a student can't write down numbers while computing, nor should students be required to write down the composition or decomposition for every problem.

In Lessons 1 through 7, mental math strategies will be reinforced and extended to three-digit numbers. For example, students will learn:

76**0** + 18**0** is also 76 tens + 18 tens	Students can then use any already known strategy to solve the problem: • 80 tens + 14 tens = 94 tens or 940 • 74 tens + 20 tens = 94 tens or 940 • 76 tens + 20 tens − 2 tens = 760 + 200 − 20 = 940
76**0** − 18**0** is also 76 tens − 18 tens	Students can then use any already known strategy to solve the problem: • 76 tens − 10 tens − 8 tens = 760 − 100 − 80 = 580 • 76 tens − 20 tens + 2 tens = 760 − 200 + 20 = 580
100 = 90 + 10 1,000 = 900 + 100	**6** 7 + **3** 3 **6** 7**0** + **3** 3**0** 9 tens 10 ones 9 hundreds 10 tens
Adding numbers close to hundreds by compensation or "over-adding"	498 + 56 = 500 + 54 = 554 500 + 56 − 2 = 554 2 54
Subtracting numbers close to hundreds by subtracting from the hundreds or "over-subtracting"	388 − 197 = 188 + 3 = 191 388 − 197 = 388 − 200 + 3 = 191 188 200

Students will continue to apply these strategies in Lessons 8 through 12 with bar models.

Encourage students to use mental strategies for problems in future chapters, pointing out mental math opportunities as they arise.

Bar Models

In **Dimensions Math® 2A**, students learned how to use a provided bar model to interpret a word problem, and to solve a word problem where the solution depended on a previously calculated step. In this chapter, students will be asked to draw the models.

Lessons 8 through 11 in this chapter further develop students understanding of bar models and extend them to multi-step problems. Students begin by modeling problems with strips of paper, then progress to drawing models.

Note: The numbers in these bar model lessons are intentionally easy for computation. The focus is on the problem solving methods.

The purpose of drawing the models is not to encourage students to follow specific rules, but to understand the concepts and choose a good problem solving method.

As students solve harder problems they will begin to appreciate the usefulness of the model drawing strategy. Students will be expected to:

* Draw a bar model for each problem.
* Write an expression for each problem.
* Solve each problem. The answer should be given in a complete sentence, for example, "Alex had $52 at first." This is to encourage students to not merely solve the problem correctly, but to solve the correct problem.

Bar models are used for word problems throughout the **Dimensions Math®** series. There are two main types of models students will work with at this level: **Part-whole Models** and **Comparison Models**.

Part-whole Models extend the understanding of part-whole relationships from a number bond to a bar model. A bar model represents pictorially the relationship between the size of the quantities more accurately.

There are **100 people** at the park. There are **40 adults** at the park. How many children are at the park?

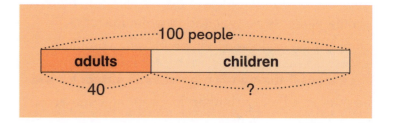

$100 - 40 = 60$ or $40 + \underline{\quad} = 100$

There are 60 children at the park.

Students can easily see from the model, as from the number bond, that since they are given the whole and one part, they subtract to find the other part.

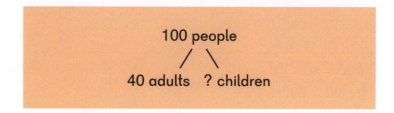

Comparison Models allow students to compare quantities. They are particularly useful in representing multi-step problems.

There are **140 boys** and **160 girls** at the park.

(a) How many more boys than girls are at the park?
(b) How many children were at the park in all?

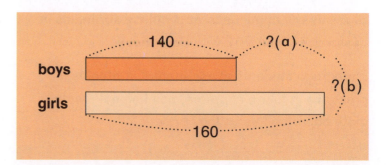

Comparison models reinforce the concepts of "more than" and "less than" and are related to the concept of difference. Students can easily see that they need to subtract to find the answer to (a). Although (b) is actually a part-whole question, the same model can be used to indicate the whole by placing a question mark at the end of the two bars.

Bar modeling is a good introduction to algebra as it provides pictorial representations of algebraic equations.

The notion of a variable can be introduced simply by labeling the units in the bar model. In the previous example, the difference between boys and girls can be labeled as "a," and the total number of boys and girls as "b."

To find the difference, students can write the equation: $160 - 140 = a$ or $140 + a = 160$

To find the total number, students can use the equation: $140 + 160 = b$

In the **Dimensions Math® 3A** textbook, students are not asked to use algebra to solve the problem, but teachers can show and explain how the bar model represents an equation.

Materials

- Art paper
- Dry erase markers
- Dry erase sleeves
- Game markers
- Two different color strips of paper of the same length for each student
- Whiteboards

Blackline Masters

- Equation Symbol Cards
- Hundred Chart
- Number Cards
- Scoring Sheet
- Sum to 1,000 Number Cards

Activities

Games and activities included in this chapter are designed to provide practice and extensions of place value concepts. They can be used after students complete the **Do** questions, or any time review and practice are needed.

Objective

- Write expressions and equations.

Have students discuss the **Chapter Opener** problems in groups. Ask them whether they should add or subtract, and why. Have some students share their thinking and represent the problems with expressions.

Dion introduces the term **expression** and contrasts it with an **equation**.

Expressions consist of numbers, operation symbols, and variables (often a box) that are grouped together to show a value or quantity. For example: 46 + 38, 24 ÷ ▢, and a ÷ 4 are expressions.

In an equation, expressions on both sides of the equal sign have the same value. 46 + 38 and 84 are both expressions. 46 + 38 = 84 is an equation.

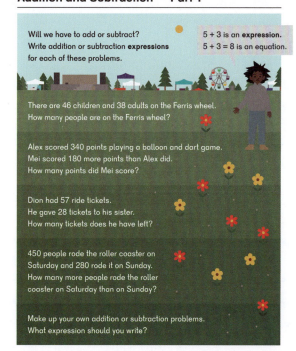

Chapter 2

Addition and Subtraction — Part 1

Will we have to add or subtract?
Write addition or subtraction **expressions** for each of these problems.

5 + 3 is an **expression**.
5 + 3 = 8 is an equation.

There are 46 children and 38 adults on the Ferris wheel. How many people are on the Ferris wheel?

Alex scored 340 points playing a balloon and dart game. Mei scored 180 more points than Alex did. How many points did Mei score?

Dion had 57 ride tickets. He gave 28 tickets to his sister. How many tickets does he have left?

450 people rode the roller coaster on Saturday and 280 rode it on Sunday. How many more people rode the roller coaster on Saturday than on Sunday?

Make up your own addition or subtraction problems. What expression should you write?

41

Lesson 1 Mental Addition — Part 1

Objective

• Learn and apply different strategies for adding two-digit numbers mentally.

Think

Pose the **Think** problem and ask students to find a solution mentally.

Have students share how they solved the problem. Record their strategies on the board and discuss the methods they used.

Learn

Have students discuss the solutions that Sofia, Mei, and Alex are presenting and have students compare their own strategies with the ones in the textbook.

Sofia is adding the tens from 38 first, then the ones.

Mei is applying the strategy of making the next 10. 38 + 2 = 40. 40 is an easy number to add to 44.

Alex notices that adding 38 to 46 is the same as adding 40 to 46 and then subtracting 2. In other words, 46 + 40 − 2 = 86 − 2 is the same as 46 + 38. This strategy is useful when the number being added is close to the next ten or hundred.

As students learn to think flexibly they will see that depending on the numbers given, certain methods may be easier to use than others.

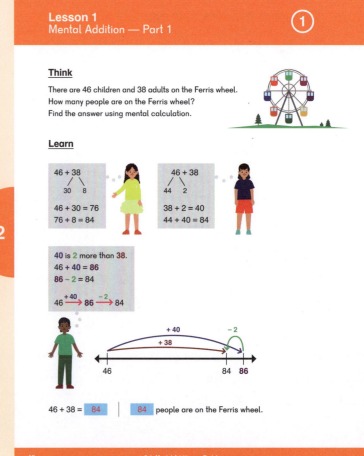

Do

Note: In this lesson no sums are greater than 100. The focus is on strategies where ones need to be regrouped, not tens and not both ones and tens.

❶ (a–d) Have students discuss alternative strategies.

(a) shows adding tens before ones, in contrast to the standard vertical algorithm. In cases where there is no regrouping, as in ❶ (a), this strategy is straightforward. ❶ (b) shows it can still work well when only one regrouping is necessary.

(b) Students could also split 56 into 5 and 51 instead of adding the tens then the ones.

(d) Since 19 is close to 20, it is easy to use the strategy of over-adding by adding 20 and then subtracting 1. Students may find it is just as easy to split 74 into 73 and 1 to make 20 with the 19. Encourage them to also think about this strategy of over-adding, as it will become useful with greater numbers.

❷ Students may use any method. These questions offer opportunities for discussion on which strategy is most efficient for each problem and each student.

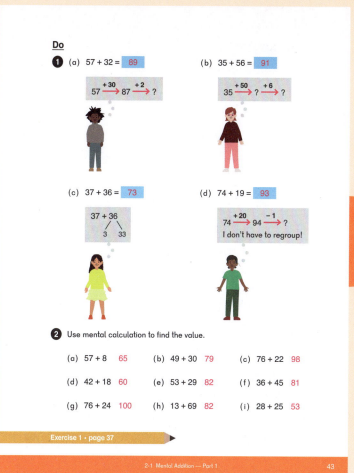

Activity

● **Three in a Row — Add**

Materials: Hundred Chart (BLM) in a dry erase sleeve, multiple sets of Number Cards (BLM) 0 to 9, dry erase markers

Shuffle the Number Cards (BLM) and deal 4 to each player. Players take turns arranging their numbers into 2 two-digit numbers to find a sum less than 100. The player marks the sum on the Hundred Chart (BLM).

If it is not possible to make two numbers whose sum is less than 100 using all 4 cards, the player may use only 3 of the cards.

The winner is the first player to mark 3 numbers in a row (horizontally, vertically, or diagonally).

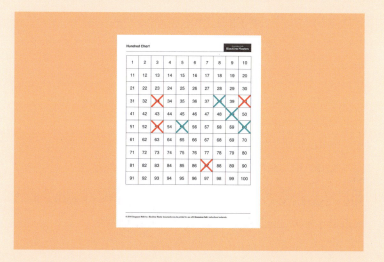

Exercise 1 • page 37

Lesson 2 Mental Addition — Part 2

Objective

- Mentally add three-digit numbers that are multiples of ten.

Think

Pose the **Think** problem and ask students to find a solution using a mental math strategy. If they are having difficulty, remind students of what they learned in the previous chapter, i.e., that 340 is the same as 34 tens, and that tens can be added to tens in the same way as ones to ones.

Have students share how they solved the problem. Record their strategies on the board and discuss the methods used.

Learn

Have students discuss the solutions that Sofia, Mei, and Alex are presenting and have students compare their own strategies with the ones in the textbook.

Sofia is splitting 180 into 100 and 80 because it is easy to add 100 to a number.

Mei is splitting 340 into 320 and 20 so she can make 200, which is easy to add. The digits highlighted in red show that she can think of the calculation as 34 tens + 18 tens and use the strategy of making the next 10.

- 180 (18 tens) needs 20 (2 tens) more to make 200, which is easy to add to 320.
- Note that students could also split the 180 to 60 and 120 to add 340 + 60 + 120.

Alex sees that adding 340 and 180 is the same as adding 340 and 200 and subtracting 20.

$$340 + 180 = 340 + 200 - 20$$

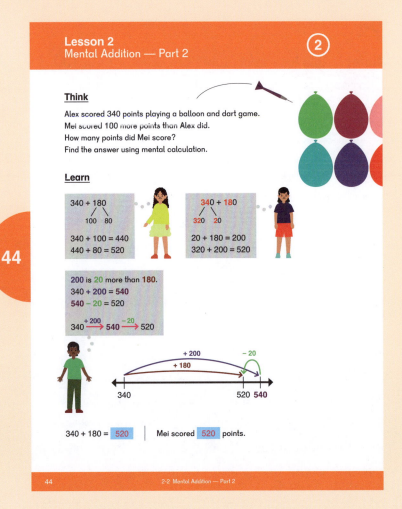

Do

1 To add 340, 300 is added first, then 40. This turns the problem into an easier problem, 780 + 40.

2 The second addend is decomposed to make the next hundred.

3 The over-adding method is easy to use because 290 is close to 300 and 70 is close to 100. The extra amount that was added has to be subtracted to get the answer.

Students may use any method for **4**. These questions also offer opportunities for discussion of which strategy is most efficient for each problem and each student.

Exercise 2 • page 41

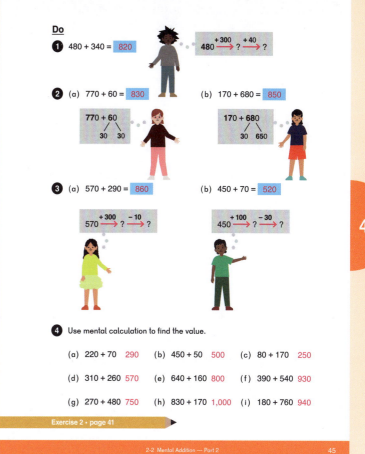

Do

1 480 + 340 = 820

480 →(+300) ? →(+40) ?

2 (a) 770 + 60 = 830

770 + 60
  ╱   ╲
 30   30

(b) 170 + 680 = 850

170 + 680
   ╱   ╲
  30   650

3 (a) 570 + 290 = 860

570 →(+300) ? →(−10) ?

(b) 450 + 70 = 520

450 →(+100) ? →(−30) ?

4 Use mental calculation to find the value.

(a) 220 + 70 290   (b) 450 + 50 500   (c) 80 + 170 250

(d) 310 + 260 570   (e) 640 + 160 800   (f) 390 + 540 930

(g) 270 + 480 750   (h) 830 + 170 1,000   (i) 180 + 760 940

Exercise 2 • page 41

2-2 Mental Addition — Part 2   45

45

Objective

- Subtract two-digit numbers mentally.

Think

Pose the **Think** problem and ask students to try finding a solution mentally.

Have students share how they solved the problem. Record their strategies on the board and discuss the methods used.

Learn

Have students discuss the solutions that Mei, Emma, and Alex are presenting and have students compare their own strategies with the ones in the textbook.

Mei is subtracting the 2 tens from 28 first, then the 8 ones.

Emma is applying the strategy of subtracting from a multiple of 10. She splits 57 into 27 and 30, then finds the difference of 30 and 28, or 2. She then adds that difference of 2 to 27.

Alex sees that subtracting 28 is the same as subtracting 30 and adding 2. That is, 57 − 28 is the same as 57 − 30 + 2.

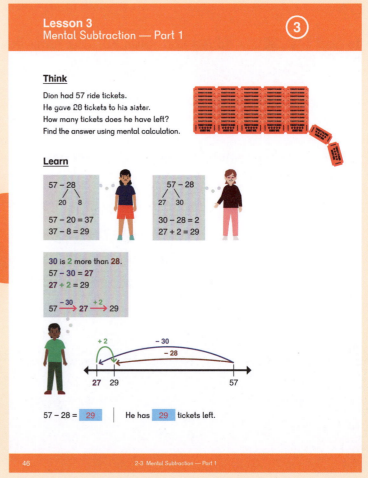

46

Do

1 (a) This problem involves no regrouping.

2 Students could also split 56 into 50 and 6 to use the strategy of double subtraction:
$81 - 50 = 31, 31 - 6 = 25$.

3 Subtracting 19 is the same as subtracting 20 and adding 1.

4 Students may use any method. These questions offer opportunities for discussion of which strategy is most efficient for each problem and each student.

Activity

● **Three in a Row — Subtract**

Materials: Hundred Chart (BLM) in a dry erase sleeve, multiple sets of Number Cards (BLM) 0 to 9, dry erase markers

Game plays similarly to **Three in a Row — Add** from Lesson 1 on page 55 in this Teacher's Guide.

Shuffle the Number Cards (BLM) and deal 4 to each player. Players take turns arranging their numbers into 2 two-digit numbers. Players find the difference of 2 two-digit numbers, and mark this on the Hundred Chart (BLM).

The winner is the first player to mark 3 numbers in a row (horizontally, vertically, or diagonally).

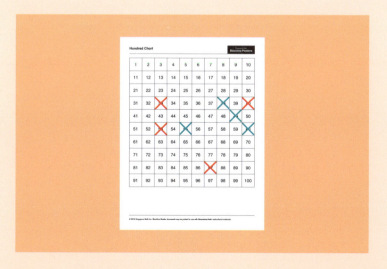

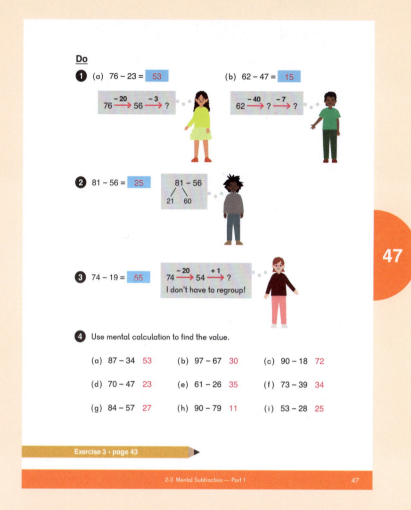

Exercise 3 • page 43

Lesson 4 Mental Subtraction — Part 2

Objective

- Mentally subtract a three-digit multiple of 10 from another three-digit multiple of 10.

Think

Pose the **Think** problem and ask students to try to find a solution mentally. Remind students what they learned in the previous chapter, i.e., that 450 is the same as 45 tens, and 280 is the same as 28 tens. Tens can be subtracted from tens in the same way as ones from ones.

Have students share how they solved the problem. Record their strategies on the board and discuss the methods used.

Learn

Have students discuss the solutions that Dion, Emma, and Mei are presenting and have students compare their own strategies with the ones in the textbook.

Dion thinks of the calculation as 45 tens − 28 tens. He chooses to break the 28 tens into 20 tens and 8 tens because it is easy to subtract 200 from 450.

Emma also thinks of 45 tens − 28 tens. She splits 450 into 150 and 300 because it is easy to subtract 28 tens from 30 tens.

- 30 tens − 28 tens = 2 tens = 20
- 20 + 150 = 170

Mei sees that subtracting 280 is the same as subtracting 300 and adding 20.

48

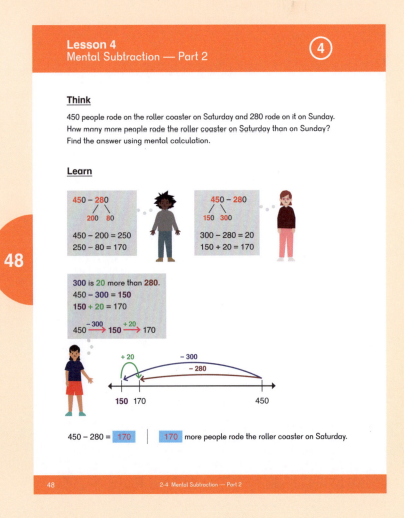

Do

1 Mei thinks of the hundreds as tens:
79 tens − 20 tens = 59 tens
59 tens − 5 tens = 54 tens = 540

2 (a) Alex splits the 50 because it is easy to subtract
2 tens from 120.
12 tens − 2 tens = 10 tens
10 tens − 3 tens = 7 tens = 70

(b) Emma splits 540 into 24 tens and 30 tens.
She can subtract 26 tens from 30 tens.
30 tens − 26 tens = 4 tens = 40
240 + 40 = 280

3 Both questions encourage students to use the
strategy of over-subtracting. Discuss with students
why this strategy was chosen for these numbers.

Exercise 4 • page 47

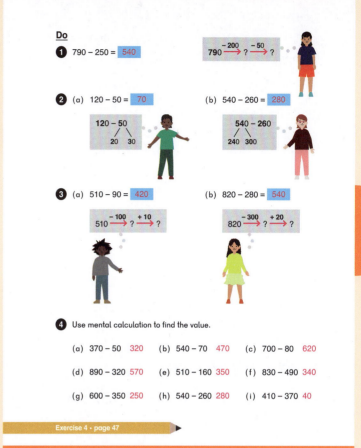

Objective

- Add or subtract a two-digit number close to a multiple of 10 or a three-digit number close to a multiple of 100 to or from a number with up to three digits.

Lesson Materials

- Number Cards (BLM) 15, 21, 36, 38, 43, 57, 62, 64, 79, 85

Think

Provide students with a set of Number Cards (BLM) with numbers from the **Think** activity.

Have them find the pairs that add up to 100 and discuss patterns they notice.

Ask students, "What do you notice about the sum of the ones and the sum of the tens in each case?"

Learn

Discuss the pairs of cards that Alex picks. Ask students for another pair of cards that makes 100 and write the solution on the board. For the cards 15 and 85, for example:

10 + 80 = 90

5 + 5 = 10

Students should note that the sum of the tens is always 9 tens when the tens are added to each other. The sum of the ones is always 10 ones.

Each pair that sums to 100 can be thought of as 90 + 10.

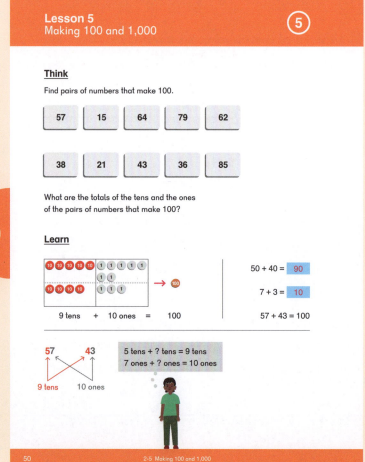

Do

❶ (a) Have students solve by asking themselves, "How many more tens will make 90?" and, "How many more ones will make 10?"

(b–d) These problems are scaffolded so that if students know 37 + 63 make 100, they can use that to apply to subtraction problems where the number of hundreds is more than 1.

100 − 37 = 63

200 − 37 = 163

Show students additional number bonds if needed:

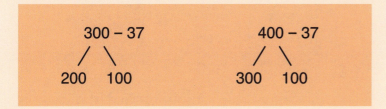

❷ This question extends the idea of making 100 to making 1,000. Students can think of 640 + 360 as 64 tens + 36 tens.

Just as 9 tens and 10 ones make 100, 9 hundreds and 10 tens makes 1,000.

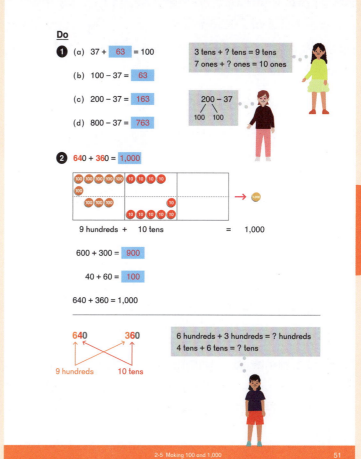

3 (a–c) are scaffolded so that if students know the number bond of 260 and 740 make 1,000, they can apply that to subtraction problems where the number of thousands is more than 1.

Show students additional number bonds if needed:

$$2,000 - 260 \qquad 3,000 - 260$$
$$\diagup\ \diagdown \qquad\qquad \diagup\ \diagdown$$
$$1,000\ \ 1,000 \qquad 2,000\ \ 1,000$$

4 Encourage students to think about the numbers as tens so they can use the same strategy they used for making 100. For example, think of 530 as 53 tens so they need to find the number that is 47 tens, which is 470.

5 (d–f) Students can split the minuend to subtract from 100. For example, in (d), split 500 into 400 and 100. 100 – 24 = 76, 76 + 400 = 476

(j–l) Students can split the minuend to subtract from 1,000. For example, in (j), split 6,000 into 5,000 and 1,000. 1,000 – 240 = 760, 760 + 5,000 = 5,760

Activity

● **Memory**

Materials: Sum to 1,000 Number Cards (BLM)

Lay Sum to 1,000 Number Cards (BLM) out in a facedown array. Players take turns turning over two cards. If the cards sum to 1,000, players have found a match and keep the cards.

If the two cards do not add up to 1,000, the cards are turned facedown again and the player's turn is over.

The player with the most cards at the end of the game is the winner.

Modify the game for students who are struggling by laying the cards faceup.

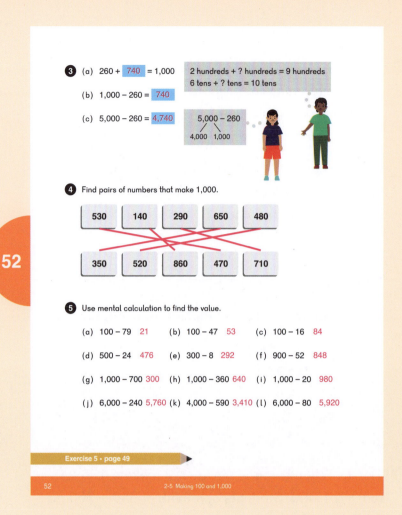

3 (a) 260 + [740] = 1,000   2 hundreds + ? hundreds = 9 hundreds
6 tens + ? tens = 10 tens

(b) 1,000 – 260 = [740]

(c) 5,000 – 260 = [4,740]   5,000 – 260
      4,000   1,000

4 Find pairs of numbers that make 1,000.

| 530 | 140 | 290 | 650 | 480 |

| 350 | 520 | 860 | 470 | 710 |

5 Use mental calculation to find the value.

(a) 100 – 79   21   (b) 100 – 47   53   (c) 100 – 16   84

(d) 500 – 24   476   (e) 300 – 8   292   (f) 900 – 52   848

(g) 1,000 – 700   300   (h) 1,000 – 360   640   (i) 1,000 – 20   980

(j) 6,000 – 240   5,760   (k) 4,000 – 590   3,410   (l) 6,000 – 80   5,920

Exercise 5 • page 49

Exercise 5 • page 49

Objective

- Add or subtract 97, 98, and 99.

Think

Pose the **Think** problem and ask the students to try to find solutions mentally.

Have students share how they solved the problem. Record their strategies on the board and discuss the methods used.

Learn

Just as students learned strategies for adding 97, 98, or 99 to a two-digit number, now students will apply those strategies for adding to a three-digit number.

Have students discuss the strategies that the four friends in the textbook used and compare them to their own methods.

Dion is solving problem (a) by making the next hundred:

98 + 2 = 100, 123 + 100 = 223.

Sofia is solving the problem by over-adding.

Note: Both methods in (a) involve subtracting 2 somewhere. Dion splits 2 from 125 and then adds 100 to 123. Sofia adds 100 first, then subtracts the 2.

Mei is solving problem (b) by subtracting from 100:

100 − 98 = 2, 25 + 2 = 27.

Emma is over-subtracting.

Note: Both methods in (b) involve adding 2 somewhere. Mei splits 25 from 125 and then subtracts 98 from 100 to get 2, 25 + 2 = 27.

Emma subtracts 100 first, then adds the 2.

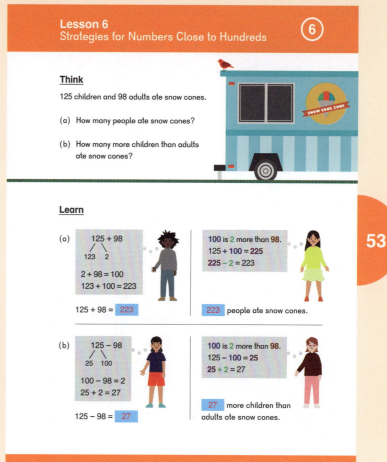

Do

① It is easy to add 400 to 526. Since we added 3 too many we need to subtract 3 to get the answer.

② It is easy to subtract 100 (or 400) from 526. Since we subtracted 3 too many we need to add 3 back to get the answer.

④ (a) Ask students if it is just as easy to add 500 and subtract 2.

(b) Both addends have 99, ask students why they think Emma chooses to split 699 instead of 199.

⑤ For the addition problems, encourage students to look for the addend that is closest to the next 100. For example, in (k), 296 is close to 300, so we can do 300 + 407 = 707, 707 − 4 = 703.

Activity

● **Add 97, 98, or 99**

Materials: Multiple decks of Number Cards (BLM) 0 to 9 and 97, 98, 99, Equation Symbol Cards (BLM)

Each player tries to make the greatest sum on a given round. Players take turns drawing four Number Cards (BLM), an Equation Symbol Card (BLM), and selecting either the 97, 98, or 99 Number Card (BLM).

On her turn, a player makes a three-digit number with one Number Card (BLM) and their 97, 98, or 99 Number Card, followed by the Equation Symbol Card (BLM). She uses the remaining Number Cards (BLM) to make a three-digit number and add the two together.

She could make 299 + 659.

In each round, the player with the total closest to 1,000 without going over gets a point. Play to 5 points.

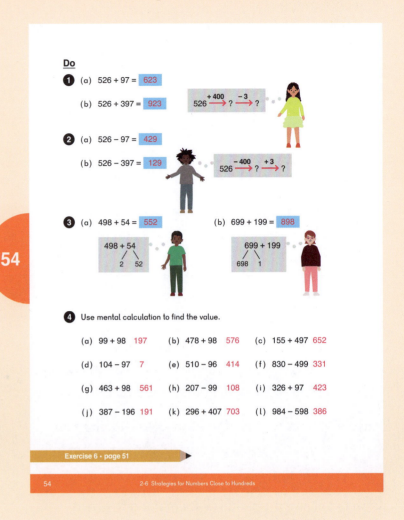

Play again, subtracting the three-digit number made with the 97, 98, or 99 card.

✏ Exercise 6 • page 51

Objective

• Practice adding and subtracting mentally.

After students complete the **Practice** in the textbook, have them continue adding numbers mentally by playing games from this chapter.

Activity

▲ **Duel**

Materials: Scoring Sheet (BLM), whiteboards and dry erase markers

Each player comes up with 3 numbers that add up to 1,000.

▲ When both players have their numbers written on their whiteboards, they take turns filling in the number duel boxes on Scoring Sheet (BLM).

The player with the greater number in each row gets a point.

For example, Player One chooses 800, 100, 100 and Player Two chooses 500, 400, 100.

Player One gets 1 point: 800 > 500. Player Two gets 1 point: 400 > 100. No points are awarded for a tie.

After 5 rounds, the player with the most points wins.

★ Students receive 1 bonus point for numbers chosen that are not multiples of one hundred.

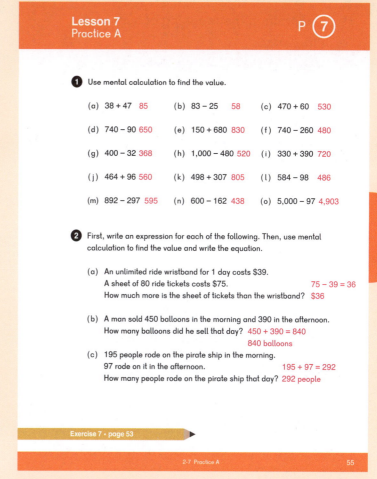

Lesson 7
Practice A   P ⑦

❶ Use mental calculation to find the value.

(a) 38 + 47   85   (b) 83 − 25   58   (c) 470 + 60   530

(d) 740 − 90 650   (e) 150 + 680 830   (f) 740 − 260 480

(g) 400 − 32 368   (h) 1,000 − 480 520   (i) 330 + 390 720

(j) 464 + 96 560   (k) 498 + 307 805   (l) 584 − 98 486

(m) 892 − 297 595   (n) 600 − 162 438   (o) 5,000 − 97 4,903

❷ First, write an expression for each of the following. Then, use mental calculation to find the value and write the equation.

(a) An unlimited ride wristband for 1 day costs $39.
A sheet of 80 ride tickets costs $75.   75 − 39 = 36
How much more is the sheet of tickets than the wristband? $36

(b) A man sold 450 balloons in the morning and 390 in the afternoon.
How many balloons did he sell that day? 450 + 390 = 840
                                         840 balloons

(c) 195 people rode on the pirate ship in the morning.
97 rode on it in the afternoon.   195 + 97 = 292
How many people rode on the pirate ship that day? 292 people

Exercise 7 • page 53

2-7 Practice A   55

55

Exercise 7 • page 53

Objectives

- Understand the meaning of sum and difference.
- Represent sums and differences using a model.

Lesson Materials

- Two different color strips of paper of the same length for each student

Think

Give each student two strips of different color paper that are the same length. Ask them what they remember about models from **Dimensions Math® 2**.

Ask students:

- How can we show the sizes of these two numbers using paper strips?
- Should the paper strips for 231 and 563 be the same length?
- How can you show both of the numbers?

Have students fold the strip that represents 231 to be shorter than the strip that represents 563.

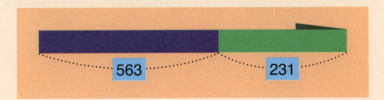

Have students discuss how to place the strips to show (a) and (b) in **Think** and ask them to share their ideas.

Ask:

- Where do we write the corresponding numbers?
- Where does the question mark belong?

Learn

Have students put the strips together, as shown in the textbook, to find the whole.

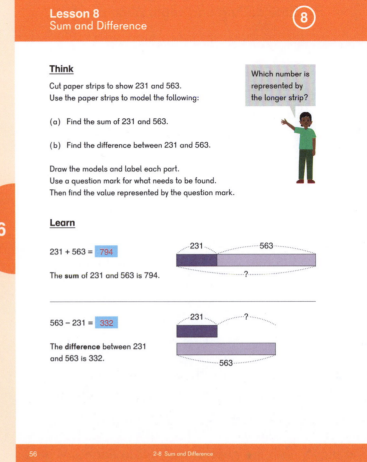

231 + 563 = 794

Introduce the term **sum** as the total or whole.

Have students place their paper strips one above the other as shown in the textbook to find the **difference**.

563 − 231 = 332

In this problem, when we subtract, we are either comparing the two numbers, or finding the difference. That is, two numbers are the same up to a certain point, then one is more than the other (by the amount known as the difference).

Do

② Mei reminds us that when we have two parts, we can add them together to find the whole.

③ Alex thinks he needs to find a missing part. Emma shows that to find the missing part, we can subtract the known part from the whole.

④ Because we are comparing the difference between 75 and 47, we can put the one strip on top of the other to compare the proportional length.

⑤ Given the difference and one number (the greater), find the other number and the sum.

Sofia thinks, "120 minus what number will make 80?" The model shows we subtract 80 from 120.

Students should know that whole − part = part, and that 120 − ? = 80 and 120 − 80 = ? will both result in the missing number.

Students find the answer to 120 − 80 = a and add that result to 120 to find the sum of the numbers for 120 + 40 = b.

Note: The model can be used to find both the difference between two numbers and the sum of the two numbers. Students are not expected to draw two different models for these problems, but may if they so choose.

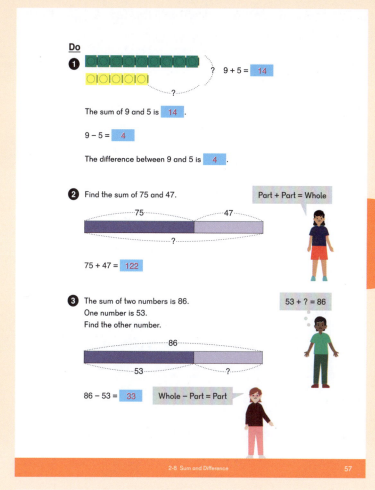

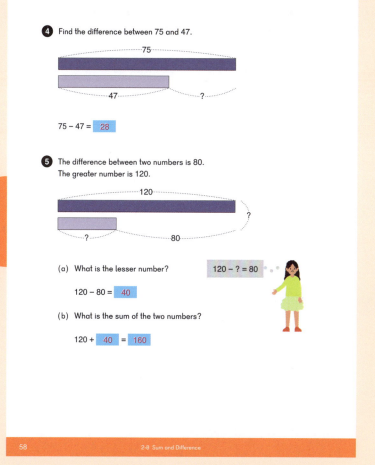

6 Given the sum and one number (the lesser), find the other number and the difference:

$$100 - 35 = a$$

Once the second number is found, we can subtract to find the difference between the two numbers:

$$65 - 35 = b$$

Note: Although the term **sum** is given in the word problem, students must subtract to find the answers.

7 Given the difference and one number (the lesser), find the other number and the sum.

$$98 + 43 = a$$

Students can then add the two numbers together to find the sum of the two numbers:

$$98 + 141 = b$$

Encourage students to use one of the mental math strategies from Lesson 6.

Note: Although the term **difference** is given in the problem, students will add to find the answers.

Exercise 8 • page 55

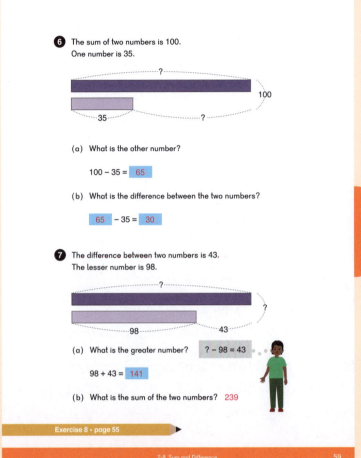

Objectives

- Understand part-whole situations in addition and subtraction.
- Represent part-whole situations using bar models to solve problems.

Think

Pose the problems from **Think**. Have students draw models to solve the problem.

Have students discuss Mei's questions. Ask:

- How are the problems the same?
- How are they different?
- How can we use what we learned in the last lesson to model this problem?
- How do the models help determine what expression to write?

Discuss student solutions.

Note: The numbers in these bar model lessons are easy to compute intentionally. The focus is on drawing representations to help students with more complex problems in the future. The task is to:

- Draw a bar model for each problem.
- Write an expression for each problem.
- Solve each problem. The answer should be given in a complete sentence. For example, "Alex has 52 tickets left."

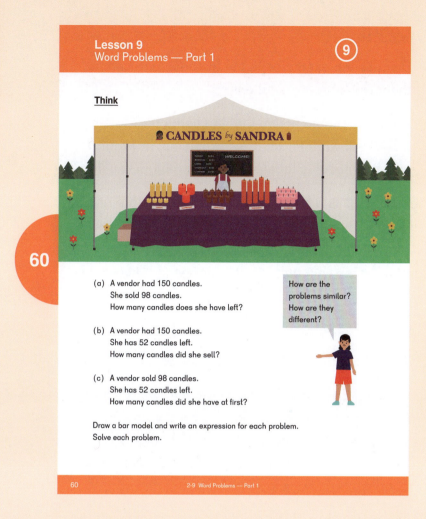

Learn

Discuss the examples in the textbook and compare them with student solutions.

Students should notice that the same part-whole model will work for all three problems, however, the way the model is labeled in each problem is different.

Problem 1 is a missing part problem. We are given the total number of candles the vendor had (the whole), and the candles sold (one part). We need to find the missing part, or the number of candles she has left:

$150 - 98 = a$

To solve, subtract the part from the whole.

Problem 2 is a missing part problem. We are given the total number of candles the vendor had (the whole), and the number of candles left (one part). We need to find the missing part, or the number of candles that she sold:

To solve, subtract the part we know (what is left) from the whole: $150 - 52 = b$

Problem 3 is a missing total problem. We are given the number of candles the vendor sold (one part) and number of candles she has left (another part). We need to find the total number of candles she had at first:

To solve, add the parts: $98 + 52 = c$

Do

❷ This model has more than two parts. Ask students where the question marks should go for each question.

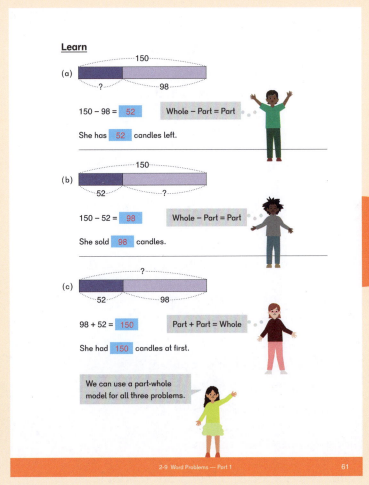

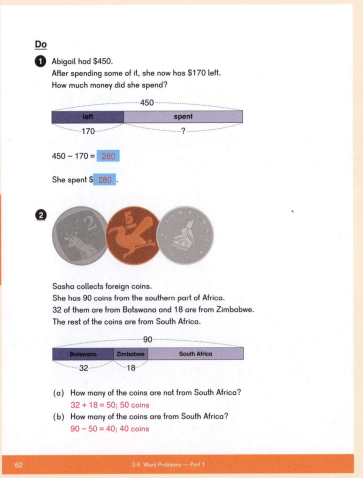

3 When students struggle with drawing models, prompt them with questions:

- What do we need to find?
- Do we know one part or two parts?
- Do we know the whole or total?

Additionally, either take the numbers out of the problems and rephrase them or use smaller numbers. This allows students to focus on the situation in the problem.

Some suggestions for removing numbers:

(a) There are some girls and some boys in the school bands. How many students are there in the bands altogether?

This shows part + part = whole.

(b) There were 640 people at the school concert. Part of the people at the concert are adults and part are children. How many children are at the concert?

Have students draw the model and label 640 as the total, then label one part as adults and the other part as children:

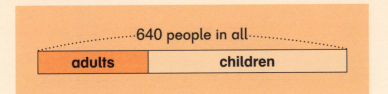

Next, give the students the number 180 for the adults. Have students label the adult part as 180 and label the children part as the question mark:

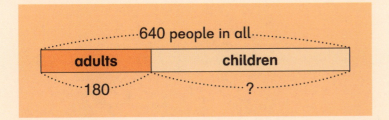

Now that they have numbers, have students estimate and change the length of the bars and solve the problem:

640 − 180 = 460

3 Draw models and solve.

(a) There are 150 girls and 160 boys in the school bands.
How many students are in the bands?
150 + 160 = 310; 310 students

(b) There are 640 people at the school concert.
180 of them are adults and the rest are children.
How many children are at the concert?
640 − 180 = 460; 460 children

(c) A bakery had 650 pretzels to sell at the start of the day.
There were 20 pretzels left at the end of the day.
How many pretzels were sold that day?
650 − 20 = 630; 630 pastries

(d) A book has 600 pages.
Jack read 430 pages.
How many pages does he still have to read?
600 − 430 = 170; 170 pages

Exercise 9 • page 60

2-9 Word Problems — Part 1   63

Exercise 9 • page 60

Objectives

- Understand comparison situations in addition and subtraction.
- Represent comparison situations using bar models to solve problems.

Think

Pose the problems from **Think**. Have students represent the quantities with paper strips or by drawing models to solve the problem.

Ask:

- How can we use what we have learned to draw a model for this problem?
- How do the models help determine what expression to write?

Discuss student solutions.

Note that all three problems use a comparison term, either "more" or "fewer," however, these key words do not indicate whether students will subtract or add. Encourage students to draw the model rather than rely on keywords.

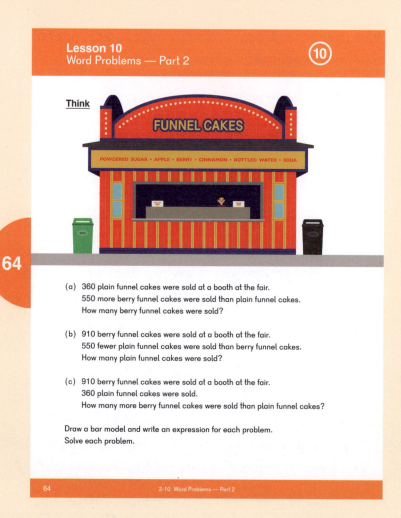

Lesson 10
Word Problems — Part 2   ⑩

Think

FUNNEL CAKES

POWDERED SUGAR • APPLE • BERRY • CINNAMON • BOTTLED WATER • SODA

64

(a) 360 plain funnel cakes were sold at a booth at the fair.
550 more berry funnel cakes were sold than plain funnel cakes.
How many berry funnel cakes were sold?

(b) 910 berry funnel cakes were sold at a booth at the fair.
550 fewer plain funnel cakes were sold than berry funnel cakes.
How many plain funnel cakes were sold?

(c) 910 berry funnel cakes were sold at a booth at the fair.
360 plain funnel cakes were sold.
How many more berry funnel cakes were sold than plain funnel cakes?

Draw a bar model and write an expression for each problem.
Solve each problem.

64   2-10 Word Problems — Part 2

Learn

Have students compare their solutions from **Think** with the ones shown in the textbook.

Students should notice that the same comparison model will work for all three problems, however, the way the model is labeled in each problem is different. Discuss how the bars are labeled and where the question mark is located in each question.

Do

These problems can be extended by asking students, "How many in all?" questions:

❶ How many people were watching the rodeo in all?

❷ How many dahlias and tulips does the flower shop have in all?

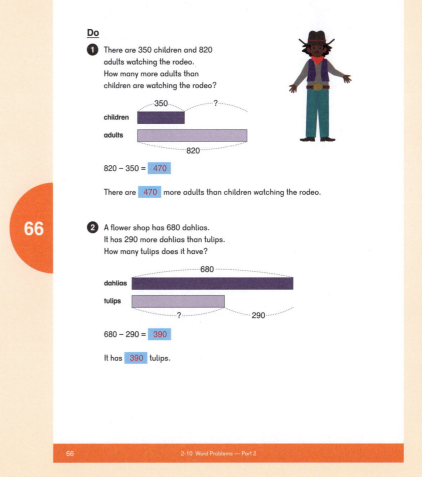

3 When students struggle with drawing models, prompt them with questions:

- What do we need to find?
- Do we know one quantity?
- Do we know how many more or fewer one quantity is?

Posing the problems without numbers works for comparison problems as well.

Some suggestions to pose problems:

(a) The flower shop has some roses and some carnations. How many more roses does it have than carnations? From the problem, we can assume that there are more roses than carnations. Students can draw:

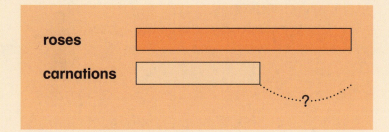

Once they have the model drawn, give students the number of roses and carnations and have them add those to their model.

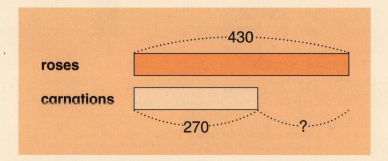

They should now see that 430 − 270 is their expression and that 430 − 270 = 160.

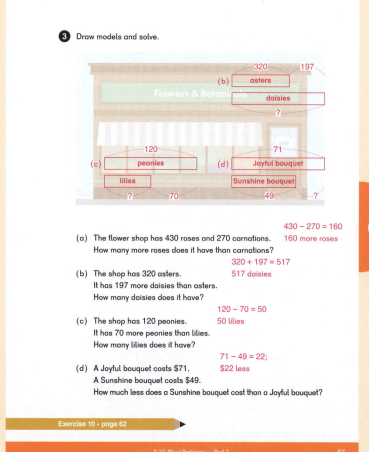

3 Draw models and solve.

(a) The flower shop has 430 roses and 270 carnations. How many more roses does it have than carnations?
430 − 270 = 160
160 more roses

(b) The shop has 320 asters. It has 197 more daisies than asters. How many daisies does it have?
320 + 197 = 517
517 daisies

(c) The shop has 120 peonies. It has 70 more peonies than lilies. How many lilies does it have?
120 − 70 = 50
50 lilies

(d) A Joyful bouquet costs $71. A Sunshine bouquet costs $49. How much less does a Sunshine bouquet cost than a Joyful bouquet?
71 − 49 = 22;
$22 less

Exercise 10 • page 62

Exercise 10 • page 62

Lesson 11 2-Step Word Problems

Objectives

- Represent multi-step problem situations with a bar model.
- Solve multi-step problems with addition and subtraction.

Think

Pose the problems from **Think**. Have students represent the quantities by drawing a model to solve the problem.

Discuss student solutions. Ask students, "What did you have to find first to solve the problem?"

Learn

Discuss the examples in the textbook and compare them with student solutions.

Ask students what information we are given in the problem:

- The number of corn dogs sold in the afternoon. (540)
- How many more corn dogs were sold in the afternoon than in the morning. (190)

Ask if they can draw a model for that information:

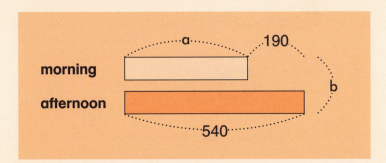

Ensure students understand the problem by asking:

- Do we know how many corn dogs were sold in the morning? (No)
- When were more sold? (Afternoon)
- If we figure out how many corn dogs were sold in the morning, can we figure out how many corn dogs were sold in all on that day? (Yes)

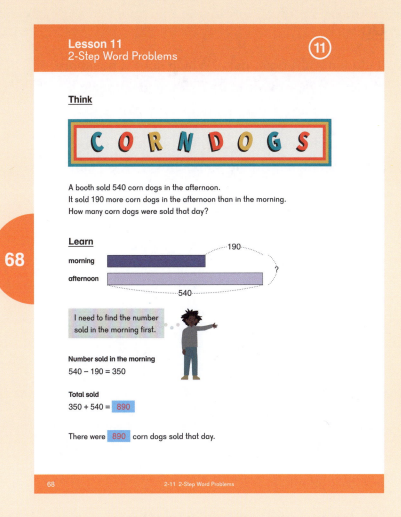

- Were 190 sold in the morning? (No)
- What part of the model shows the quantity the question is asking for?

Have students label the bar that represents the morning with a question mark, and add a question mark to the right (b) to show how many corn dogs were sold altogether.

Do

1 If students struggle, ask, "Which school has more students? How many more?" Remind students that if one school has fewer, the other school has more.

2 In Lesson 9 (textbook page 60), students solved a similar problem with the steps broken down for them.

In this problem, Sofia first subtracts the number of students in class 3A from the total number of third graders. She then subtracts the number of students in class 3C from that number to find how many students are in class 3B.

Mei adds together the number of students in the 3A and 3C classes first, and then subtracts that from the total number of third graders to find the number of students in the 3B class.

Both solutions can be found using the same model.

An answer in a complete sentence would be, "There are 23 students in 3B."

In these problems, only the information given is labeled, and what students need to find is not.

3 If students are struggling, suggest that they redraw the model themselves to help see the difference in the quantities represented by the Upper School and Lower School bars.

4 To find how much money Ivy saved, we first need to find how much Emily saved. Ask students, "How can we determine how much money Emily saved?"

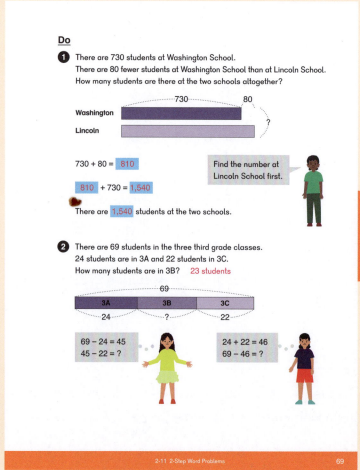

5 Prompt students with questions:

- What do we know? Are we trying to find parts or wholes, or are we comparing?
- How many bars should we draw?
- What are we trying to find? Can we label that?

Ask:

(a)
- What must we know first to answer the question? (The number of students in grade 5)
- Do we know that already? How can we find it?

(b)
- Which bar is longer? How much longer?

(c)
- What must we know to find out how much less was made selling zucchini muffins?
- How can we find out?

(d)
- What must we first determine to know how much money Ella has?

Exercise 11 • page 64

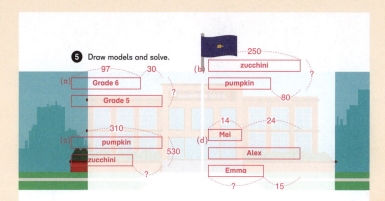

5 Draw models and solve.

(a) 250 zucchini 97 30 Grade 6 pumpkin Grade 5 ? 80 ?

(c) 310 pumpkin (d) 14 24 Mei zucchini Alex 530 Emma ? 15

(a) 97 students are in grade 6 at Washington School.   97 + 30 = 127
30 more students are in grade 5 than grade 6.   97 + 127 = 224
How many students are in both grades?   224 students

(b) The students in a cooking club baked 250 zucchini muffins.   250 − 80 = 170
They baked 80 fewer pumpkin muffins than zucchini muffins   250 + 170 = 420
How many muffins did they bake in all?   420 muffins

(c) The school made $530 at a bake sale.   530 − 310 = 220
$310 came from selling pumpkin muffins.   310 − 220 = 90
The rest came from selling zucchini muffins. $90 less
How much less was made from selling zucchini
muffins than pumpkin muffins?

(d) Maya sold 14 muffins at the bake sale.   14 + 24 = 38
Adam sold 24 more muffins than Maya.   38 − 15 = 23
Ella sold 15 fewer muffins than Adam.   23 muffins
How many muffins did Ella sell?

Exercise 11 • page 64

Objective

- Practice skills and concepts from the chapter.

Exercise 12 • page 68

Brain Works

★ Sprouts

Materials: Art paper or whiteboards

Players begin with 2 points on a paper and take turns drawing lines and points by these rules:

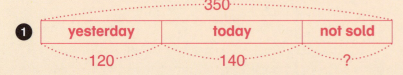

- A line must start and end on a point. (Loops are allowed):

- Once a player draws a line, he marks a new point somewhere on the line:

- The next player draws his line and places a point on it:
- Lines may not cross.

- Each point can have no more than 3 lines coming from it.
- The last person able to draw a line is the winner.

Lesson 12
Practice B

P ⑫

① Cooper has 350 raffle tickets to sell.
He sold 120 tickets yesterday and 140 tickets today.
How many raffle tickets has he not sold?
120 + 140 = 260; 350 − 260 = 90; 90

② Bayla spent $470 on a game console.
She spent $280 less on a controller than on the game console.
How much did she spend in all?
470 − 280 = 190; 470 + 190 = 660; $660

③ A guitar costs $50 more than a banjo.
The banjo costs $140.
How much do both instruments cost altogether?
140 + 50 = 190; 140 + 190 = 330; $330

④ The difference between two numbers is 90.
The greater number is 310.
What is the sum of the two numbers?
310 − 90 = 220; 310 + 220 = 530; 530

⑤ The difference between two numbers is 30.
The lesser number is 90.
What is the sum of the two numbers?
90 + 30 = 120; 120 + 90 = 210; 210

⑥ The sum of two numbers is 110.
One of the numbers is 50.
What is the difference between the two numbers?
110 − 50 = 60; 60 − 50 = 10; 10

72

Exercise 12 • page 68

72   2-12 Practice B

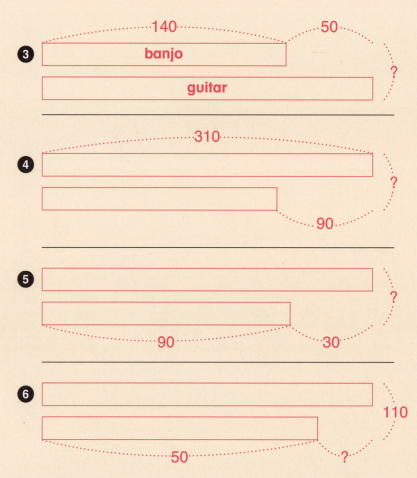

Chapter 2 Addition and Subtraction — Part 1

Exercise 1

Basics

1 (a) Add 6 to 78 by making the next ten.

$$78 + 6 = \boxed{80} + 4 = \boxed{84}$$
$$\underset{2}{\diagdown} \quad \underset{4}{\diagdown}$$

(b) Add 48 and 36 by adding tens and then ones.

$$48 \xrightarrow{+30} \boxed{78} \xrightarrow{+6} \boxed{84}$$

2 Add 27 and 45 by making the next ten.

$$27 + 45 = \boxed{30} + 42 = \boxed{72}$$
$$\underset{3}{\diagdown} \quad \underset{42}{\diagdown}$$

3 Add 54 and 38 by adding 40 and subtracting 2.

$$54 \xrightarrow{+40} \boxed{94} \xrightarrow{-2} \boxed{92}$$

Practice

4 (a) $49 + 6 = \boxed{55}$ (b) $35 + 7 = \boxed{42}$
$\quad\quad\ \underset{1}{\diagdown}\ \underset{5}{}$ $\quad\quad\quad\quad \underset{5}{\diagdown}\ \underset{2}{}$

(c) $46 + 5 = \boxed{51}$ (d) $28 + 4 = \boxed{32}$
$\quad \underset{4}{\diagdown}\ \underset{1}{}$ $\quad\quad\quad \underset{2}{\diagdown}\ \underset{2}{}$

(e) $63 + 20 = \boxed{83}$ (f) $28 + 70 = \boxed{98}$

5 (a) $67 \xrightarrow{+20} \boxed{87} \xrightarrow{+7} \boxed{94}$

$67 + 27 = \boxed{94}$

(b) $23 \xrightarrow{+50} \boxed{73} \xrightarrow{+8} \boxed{81}$

$23 + 58 = \boxed{81}$

(c) $16 \xrightarrow{+40} \boxed{56} \xrightarrow{+5} \boxed{61}$

$16 + 45 = \boxed{61}$

6 (a) $57 + \boxed{3} = 60$ (b) $86 + \boxed{4} = 90$

(c) $23 + \boxed{7} = 30$ (d) $74 + \boxed{6} = 80$

(e) $48 - 2 = \boxed{46}$ (f) $97 - 4 = \boxed{93}$

7 (a) $38 + 33 = \boxed{71}$
$\quad\quad \underset{2}{\diagdown}\ \boxed{31}$

(b) $17 + 49 = \boxed{66}$
$\quad \underset{3}{\diagdown}\ \underset{46}{}$

(c) $67 + 23 = \boxed{90}$
$\quad\quad \underset{3}{\diagdown}\ \boxed{20}$

(d) $55 + 18 = \boxed{73}$
$\quad\quad \underset{5}{\diagdown}\ \boxed{13}$

(e) $23 + 68 = \boxed{91}$
$\quad \boxed{21}\ \underset{2}{\diagdown}$

8 Find the value.

(a) $36 + 59$

$$36 \xrightarrow{+60} \boxed{96} \xrightarrow{-1} \boxed{95}$$

(b) $23 + 48$

$$23 \xrightarrow{+50} \boxed{73} \xrightarrow{-2} \boxed{71}$$

(c) $65 + 17$

$$65 \xrightarrow{+20} \boxed{85} \xrightarrow{-3} \boxed{82}$$

Challenge

9 Use mental calculation to find the value.

(a) $67 + 85 = \boxed{152}$ (b) $58 + 74 = \boxed{132}$

(c) $77 + 89 = \boxed{166}$ (d) $46 + 88 = \boxed{134}$

(e) $86 + 57 = \boxed{143}$ (f) $75 + 59 = \boxed{134}$

Teacher's Guide 3A Chapter 2

Exercise 2

Basics

1 (a) Write the missing numbers.

$$780 + 60 = \boxed{800} + 40 = \boxed{840}$$
$$\underset{20 \quad 40}{\diagdown}$$

(b) Add 480 and 360 by adding hundreds and then tens.

$$480 \xrightarrow{+300} \boxed{780} \xrightarrow{+60} \boxed{840}$$

2 Add 270 and 450 by making the next hundred.

$$270 + 450 = \boxed{300} + 420 = \boxed{720}$$
$$\underset{30 \quad 420}{\diagdown}$$

3 Add 540 and 380 by adding 400 and subtracting 20.

$$540 \xrightarrow{+400} \boxed{940} \xrightarrow{-20} \boxed{920}$$

4 Since 34 + 27 = __61__, 34 tens + 27 tens = __61__ tens.

Practice

5 Add.

390 + 250 = **640** **N**	220 + 480 = **700** **T**	570 + 360 = **930** **I**
670 + 150 = **820** **E**	750 + 160 = **910** **A**	520 + 290 = **810** **N**
480 + 390 = **870** **O**	370 + 170 = **540** **T**	240 + 280 = **520** **R**
250 + 480 = **730** **P**	660 + 260 = **920** **U**	380 + 380 = **760** **G**

Each person has a different finger print.
What other type of print is different in each person?
Write the letters to match the answers above to find out.

A		T	O	N	G	U	E
910	830	540	870	640	760	920	820

	P	R	I	N	T		
600	730	520	930	810	700	750	780

Teacher's Guide 3A Chapter 2

Exercise 3

Basics

1 (a) Subtract 6 from 43 by subtracting 6 from 40.

$$43 - 6 = 3 + \boxed{34} = \boxed{37}$$
3   40

(b) Subtract 46 from 83 by subtracting tens and then ones.

83 $\xrightarrow{-40}$ $\boxed{43}$ $\xrightarrow{-6}$ $\boxed{37}$

2 Subtract 27 from 75 by subtracting 27 from 30.

$$75 - 27 = 45 + \boxed{3} = \boxed{48}$$
45   30

3 Subtract 28 from 86 by subtracting 30 and adding 2.

86 $\xrightarrow{-30}$ $\boxed{56}$ $\xrightarrow{+2}$ $\boxed{58}$

Practice

4 (a) 34 − 7 = $\boxed{27}$ (b) 34 − 7 = $\boxed{27}$
4   30   4   3

(c) 82 − 6 = $\boxed{76}$ (d) 22 − 5 = $\boxed{17}$

(e) 63 − 20 = $\boxed{43}$ (f) 78 − 30 = $\boxed{48}$

5 (a) 64 $\xrightarrow{-20}$ $\boxed{44}$ $\xrightarrow{-6}$ $\boxed{38}$

64 − 26 = $\boxed{38}$

(b) 73 $\xrightarrow{-50}$ $\boxed{23}$ $\xrightarrow{-8}$ $\boxed{15}$

73 − 58 = $\boxed{15}$

(c) 82 $\xrightarrow{-50}$ $\boxed{32}$ $\xrightarrow{-4}$ $\boxed{28}$

82 − 54 = $\boxed{28}$

6 (a) 80 − 72 = $\boxed{8}$ (b) 90 − 88 = $\boxed{2}$

(c) 40 − 31 = $\boxed{9}$ (d) 30 − 27 = $\boxed{3}$

(e) 63 = 70 − $\boxed{7}$ (f) 45 = 50 − $\boxed{5}$

(g) 82 − 60 = $\boxed{22}$ (h) 38 − $\boxed{30}$ = 8

7 Solve by subtracting from tens.

(a) 73 − 35 = $\boxed{38}$ (b) 64 − 48 = $\boxed{16}$
$\boxed{33}$ 40 $\boxed{14}$ 50

(c) 75 − 36 = $\boxed{39}$ (d) 81 − 67 = $\boxed{14}$
35 $\boxed{40}$ 11 $\boxed{70}$

(e) 94 − 27 = $\boxed{67}$ (f) 21 − 15 = $\boxed{6}$
$\boxed{64}$ 30 $\boxed{1}$ 20

8 Find the value.

(a) 82 − 59

82 $\xrightarrow{-60}$ $\boxed{22}$ $\xrightarrow{+1}$ $\boxed{23}$

(b) 75 − 48

75 $\xrightarrow{-50}$ $\boxed{25}$ $\xrightarrow{+2}$ $\boxed{27}$

(c) 66 − 17 = $\boxed{49}$

66 $\xrightarrow{-20}$ $\boxed{46}$ $\xrightarrow{+3}$ $\boxed{49}$

Challenge

9 Use mental calculation to find the value.

(a) 374 − 59 = $\boxed{315}$ (b) 483 − 28 = $\boxed{455}$

(c) 365 − 28 = $\boxed{337}$ (d) 290 − 77 = $\boxed{213}$

(e) 887 − 48 = $\boxed{839}$ (f) 782 − 59 = $\boxed{723}$

Teacher's Guide 3A Chapter 2

Exercise 4

Basics

1 (a) Subtract 60 from 430 by subtracting 60 from 400.

$430 - 60 = 30 + \boxed{340} = \boxed{370}$

30 400

(b) Subtract 460 from 830 by subtracting hundreds and then tens.

$830 \xrightarrow{-400} \boxed{430} \xrightarrow{-60} \boxed{370}$

2 Subtract 270 from 750 by subtracting 270 from 300.

$750 - 270 = 450 + \boxed{30} = \boxed{480}$

450 300

3 Subtract 280 from 860 by subtracting 300 and adding 20.

$860 \xrightarrow{-300} \boxed{560} \xrightarrow{+20} \boxed{580}$

4 Since $82 - 27 = \underline{55}$, 82 tens – 27 tens = $\underline{55}$ tens.

Practice

5 Subtract.

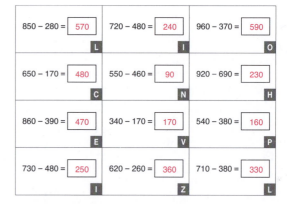

$850 - 280 = \boxed{570}$ **L**	$720 - 480 = \boxed{240}$ **I**	$960 - 370 = \boxed{590}$ **O**
$650 - 170 = \boxed{480}$ **C**	$550 - 460 = \boxed{90}$ **N**	$920 - 690 = \boxed{230}$ **H**
$860 - 390 = \boxed{470}$ **E**	$340 - 170 = \boxed{170}$ **V**	$540 - 380 = \boxed{160}$ **P**
$730 - 480 = \boxed{250}$ **I**	$620 - 260 = \boxed{360}$ **Z**	$710 - 380 = \boxed{330}$ **L**

What are some other words for ZERO?
Write the letters to match the answers above to find out.

Z	I	L	C	H		N	I	L
360	250	330	480	230	150	90	240	570

L	O	V	E		Z	I	P	
330	590	170	470	80	360	250	160	340

Exercise 5

Basics

1 $100 = 9$ tens + $\boxed{10}$ ones

$1,000 = 9$ hundreds + $\boxed{10}$ tens

2 (a) $50 + \boxed{40} = 90$

$7 + \boxed{3} = 10$

$57 + \boxed{43} = 100$

(b) $500 + \boxed{400} = 900$

$70 + \boxed{30} = 100$

$570 + \boxed{430} = 1,000$

Practice

3 Match numbers that make 1,000.

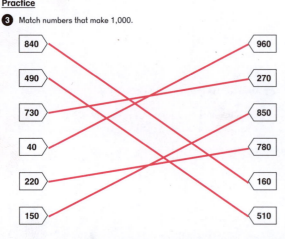

840
490
730
40
220
150

960
270
850
780
160
510

4 (a) $82 + \boxed{18} = 100$

(b) $630 + \boxed{370} = 1,000$

(c) $220 + \boxed{780} = 1,000$

(d) $34 + \boxed{66} = 100$

(e) $30 + \boxed{970} = 1,000$

(f) $8 + \boxed{92} = 100$

(g) $\boxed{56} + 44 = 100$

(h) $\boxed{170} + 830 = 1,000$

5 (a) $100 - 69 = \boxed{31}$

(b) $1,000 - 490 = \boxed{510}$

(c) $1,000 - 520 = \boxed{480}$

(d) $1,000 - 70 = \boxed{930}$

(e) $100 - 18 = \boxed{82}$

(f) $1,000 - 250 = \boxed{750}$

(g) $100 - \boxed{42} = 58$

(h) $1,000 - \boxed{830} = 170$

Challenge

6 $1,000 = 9$ hundreds + 9 tens + $\boxed{10}$ ones

7 Use mental calculation to find the value.

(a) $637 + \boxed{363} = 1,000$

(b) $142 + \boxed{858} = 1,000$

(c) $1,000 - 42 = \boxed{958}$

(d) $1,000 - 7 = \boxed{993}$

(e) $1,000 - 333 = \boxed{667}$

(f) $1,000 - 307 = \boxed{693}$

Exercise 6

Basics

1

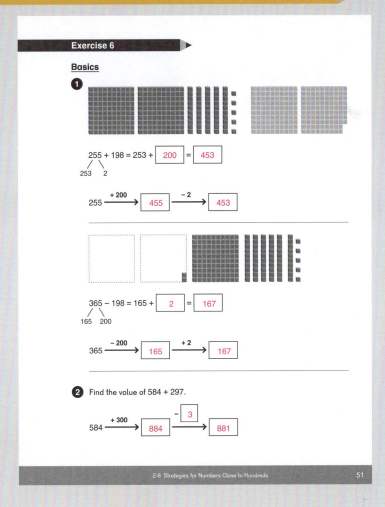

$255 + 198 = 253 +$ [200] $=$ [453]

253 2

$255 \xrightarrow{+\ 200}$ [455] $\xrightarrow{-\ 2}$ [453]

$365 - 198 = 165 +$ [2] $=$ [167]

165 200

$365 \xrightarrow{-\ 200}$ [165] $\xrightarrow{+\ 2}$ [167]

2 Find the value of $584 + 297$.

$584 \xrightarrow{+\ 300}$ [884] $\xrightarrow{-\ 3}$ [881]

3 Find the value of $937 - 599$.

$937 \xrightarrow{-\ 600}$ [337] $\xrightarrow{+\ 1}$ [338]

Practice

4 (a) $174 + 98 =$ [272]　　(b) $174 - 98 =$ [76]

(c) $532 - 397 =$ [135]　　(d) $532 + 397 =$ [929]

(e) $498 + 299 =$ [797]　　(f) $498 - 299 =$ [199]

(g) $555 + 96 =$ [651]　　(h) $555 - 96 =$ [459]

(i) $343 - 299 =$ [44]　　(j) $343 + 299 =$ [642]

5 (a) $397 + 425 =$ [822]　　(b) $954 - 497 =$ [457]

(c) [635] $- 499 = 136$　　(d) [625] $+ 298 = 923$

(e) $328 +$ [98] $= 426$　　(f) $832 -$ [99] $= 733$

Challenge

6 Use mental calculation to find the value.

(a) $999 + 99 + 9 =$ [1,107]　　(b) $97 + 998 + 6 =$ [1,101]

(c) $64 + 39 + 99 =$ [202]　　(d) $598 + 9 + 59 =$ [666]

(e) $1,782 + 990 =$ [2,772]　　(f) $7,897 + 960 =$ [8,857]

Exercise 7

Check

1 Add or subtract.

$450 + 170 =$ [620]	$861 - 98 =$ [763]	$630 - 60 =$ [570]
$168 + 7 =$ [175]	$800 - 47 =$ [753]	$370 + 580 =$ [950]
$760 - 490 =$ [270]	$290 + 40 =$ [330]	$1,000 - 460 =$ [540]
$432 + 97 =$ [529]	$342 + 427 =$ [769]	$400 - 32 =$ [368]
$868 - 347 =$ [521]	$332 - 8 =$ [324]	$614 + 99 =$ [713]
$34 + 66 =$ [100]	$825 - 96 =$ [729]	$1,000 - 940 =$ [60]

If you add the two digits of this number together, then double that answer, you get the number again.
What number is it?
Color the boxes that match the answers you found above to find out.

80	200	300	215	735	327	450
314	529	519	620	540	324	890
610	713	940	950	42	763	300
869	270	925	100	753	521	712
42	570	70	769	773	60	768
629	729	260	368	175	330	507
421	370	199	201	303	898	999

2 Write >, <, or = in the ◯.

36 tens $+ 42$ tens ◯= 350 ones $+ 430$ ones

$350 + 640$ ◯> $120 + 740$

$580 - 120$ ◯< $320 + 190$

$530 + 150$ ◯> $930 - 270$

$670 + 350$ ◯= $650 + 370$

$2,000 - 1,420$ ◯< $1,000 - 240$

$347 + 98$ ◯= $543 - 98$

Challenge

3 Put the numbers 30, 50, 70, 90, 110, 130, 150, 170, and 190 in the magic square so that the sum of the numbers in any row, column, or diagonal is 330.

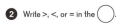

170	30	130
70	110	150
90	190	50

Hint: Start with the middle number when the numbers are listed in order in the middle square.

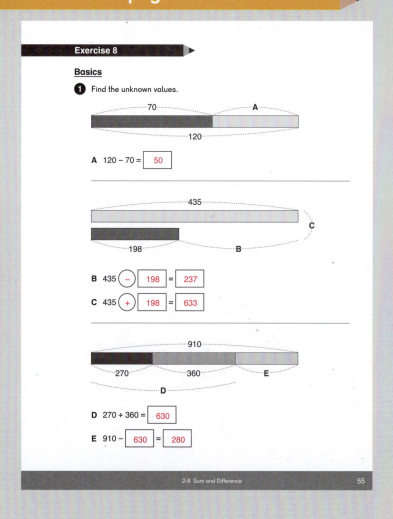

Exercise 8

Basics

1 Find the unknown values.

A 120 − 70 = 50

B 435 (−) 198 = 237

C 435 (+) 198 = 633

D 270 + 360 = 630

E 910 − 630 = 280

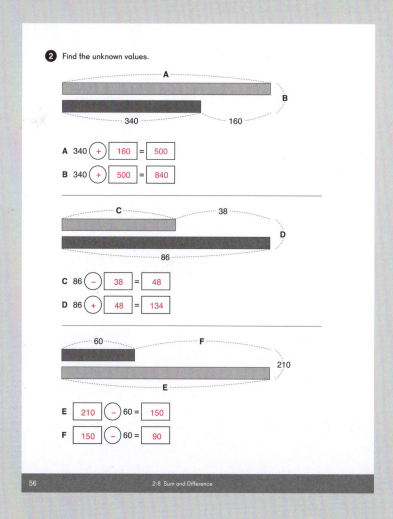

2 Find the unknown values.

A 340 (+) 160 = 500

B 340 (+) 500 = 840

C 86 (−) 38 = 48

D 86 (+) 48 = 134

E 210 (−) 60 = 150

F 150 (−) 60 = 90

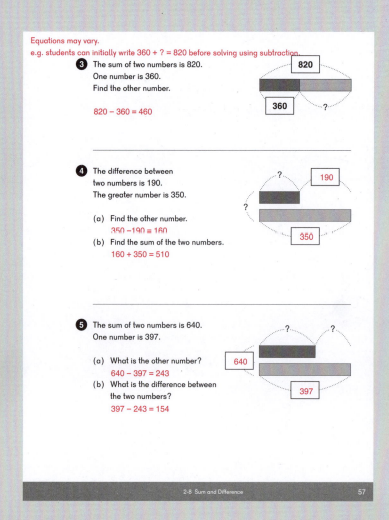

Equations may vary.
e.g. students can initially write 360 + ? = 820 before solving using subtraction.

3 The sum of two numbers is 820.
One number is 360.
Find the other number.

820 − 360 = 460

4 The difference between
two numbers is 190.
The greater number is 350.

(a) Find the other number.
350 − 190 = 160
(b) Find the sum of the two numbers.
160 + 350 = 510

5 The sum of two numbers is 640.
One number is 397.

(a) What is the other number?
640 − 397 = 243
(b) What is the difference between
the two numbers?
397 − 243 = 154

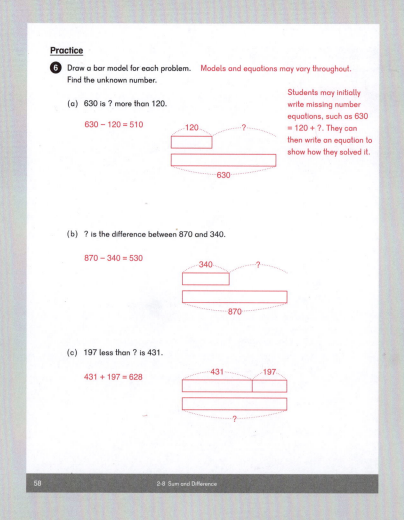

Practice

6 Draw a bar model for each problem.   *Models and equations may vary throughout.*
Find the unknown number.

(a) 630 is ? more than 120.

630 − 120 = 510

*Students may initially
write missing number
equations, such as 630
= 120 + ?. They can
then write an equation to
show how they solved it.*

(b) ? is the difference between 870 and 340.

870 − 340 = 530

(c) 197 less than ? is 431.

431 + 197 = 628

(d) The sum of 80 and ? is 210.

$210 - 80 = 130$

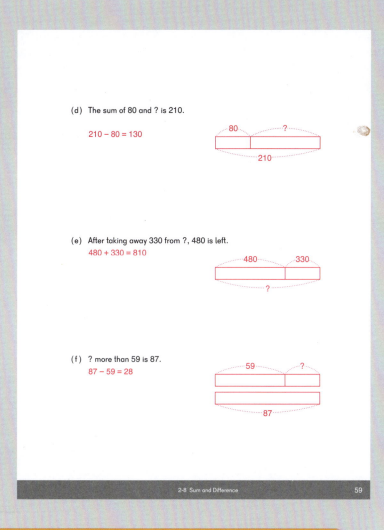

(e) After taking away 330 from ?, 480 is left.

$480 + 330 = 810$

(f) ? more than 59 is 87.

$87 - 59 = 28$

Exercise 9 • pages 60–61

Exercise 9

Basics

1 Label the bar models with the information given in the problem.
Write an expression and then solve.

(a) Pablo sold 122 tickets in the morning and 346 tickets in the afternoon.
How many tickets did he sell that day?

$122 + 346 = 468$

He sold ___468___ tickets that day.

(b) After selling 346 tickets, Katherine had 122 tickets left.
How many tickets did she have at first?

$122 + 346 = 468$

She had ___468___ tickets at first.

(c) Wyatt had 346 tickets. After selling some tickets, he had 122 tickets left.
How many tickets did he sell?

$346 - 122 = 224$

He sold ___224___ tickets.

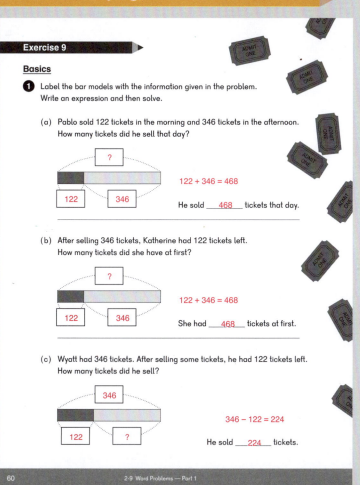

Practice

Draw bar models and solve. Models may vary.
Exact wording in answer sentences may vary.
Methods may vary.

2 A bakery has 920 chocolate chip cookies to sell.
360 of them have nuts, and the rest do not.
How many cookies did not have nuts?

| nuts | no nuts |

$920 - 360 = 560$
560 cookies did not have nuts.

3 The bakery sold 830 pastries one day.
At the end of the day, it had 80 pastries left.
How many pastries did it have at the start of the day?

| sold | left |

$830 + 80 = 910$
The bakery had 910 pastries at the start of the day.

4 The bakery sold 820 twists.
310 of them were savory twists, and the rest were cinnamon twists.
Of the cinnamon twists, 290 had nuts and the rest did not.

(a) How many cinnamon twists did the bakery sell?

(b) How many of cinnamon twists did not have nuts?

| savory | nuts | no nuts |

$820 - 310 = 510$
The bakery sold 510 cinnamon twists.

$510 - 290 = 220$
220 cinnamon twists did not have nuts.

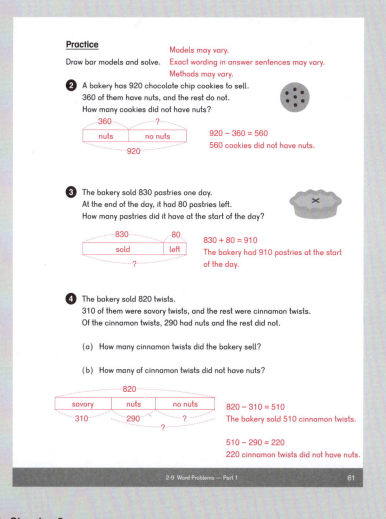

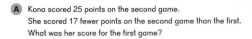

Exercise 10

Basics

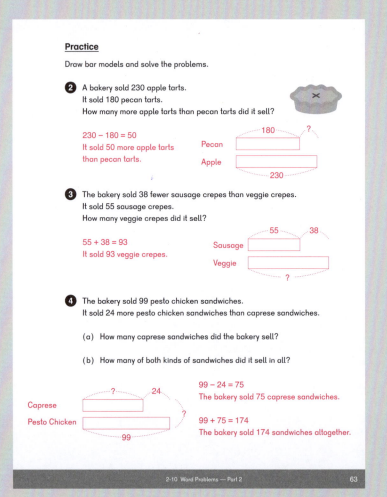

1 Which bar model goes with each problem?

A Kona scored 25 points on the second game.
She scored 17 fewer points on the second game than the first.
What was her score for the first game?

B Kona scored 17 fewer points on the second game than the first game.
She scored 25 points on the first game.
How much did she score on the second game?

C Kona scored 25 points on both games.
She scored 17 points on the second game.
What was her score for the first game?

D Kona scored 17 points on the first game.
She scored 25 points on the second game.
How many fewer points did she score on the first game than the second game?

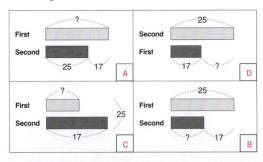

Practice

Draw bar models and solve the problems.

2 A bakery sold 230 apple tarts.
It sold 180 pecan tarts.
How many more apple tarts than pecan tarts did it sell?

230 − 180 = 50
It sold 50 more apple tarts
than pecan tarts.

Pecan

Apple

180   ?

230

3 The bakery sold 38 fewer sausage crepes than veggie crepes.
It sold 55 sausage crepes.
How many veggie crepes did it sell?

55 + 38 = 93
It sold 93 veggie crepes.

Sausage

Veggie

55   38

?

4 The bakery sold 99 pesto chicken sandwiches.
It sold 24 more pesto chicken sandwiches than caprese sandwiches.

(a) How many caprese sandwiches did the bakery sell?

(b) How many of both kinds of sandwiches did it sell in all?

Caprese

Pesto Chicken

?   24

?

99

99 − 24 = 75
The bakery sold 75 caprese sandwiches.

99 + 75 = 174
The bakery sold 174 sandwiches altogether.

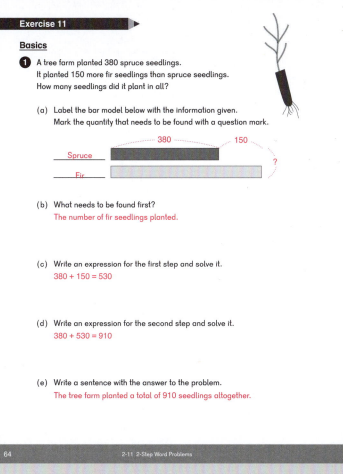

Exercise 11

Basics

1 A tree farm planted 380 spruce seedlings.
It planted 150 more fir seedlings than spruce seedlings.
How many seedlings did it plant in all?

(a) Label the bar model below with the information given.
Mark the quantity that needs to be found with a question mark.

Spruce ___ 380 ___ 150
Fir ___ ?

(b) What needs to be found first?
The number of fir seedlings planted.

(c) Write an expression for the first step and solve it.
380 + 150 = 530

(d) Write an expression for the second step and solve it.
380 + 530 = 910

(e) Write a sentence with the answer to the problem.
The tree farm planted a total of 910 seedlings altogether.

Practice

Label the bar models with the information in the problems and solve the problems.

2 A chain saw, pruner, and tree shovel cost $240 altogether.
The pruner costs $60 and the shovel costs $50.
How much does the chain saw cost?

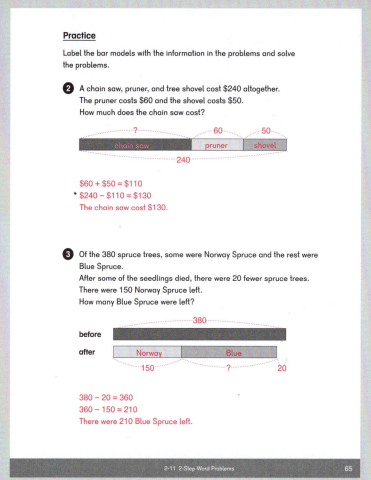

? | 60 | 50
chain saw | pruner | shovel
240

$60 + $50 = $110
$240 − $110 = $130
The chain saw cost $130.

3 Of the 380 spruce trees, some were Norway Spruce and the rest were Blue Spruce.
After some of the seedlings died, there were 20 fewer spruce trees.
There were 150 Norway Spruce left.
How many Blue Spruce were left?

before | 380
after | Norway | Blue
150 | ? | 20

380 − 20 = 360
360 − 150 = 210
There were 210 Blue Spruce left.

4 The tree lot had 98 Fraser firs for sale.
It had 36 more Douglas firs than Fraser firs for sale.
It had 12 fewer Balsam firs than Douglas firs for sale.

(a) How many Balsam firs did it have for sale?

(b) How many firs did it have for sale in all?

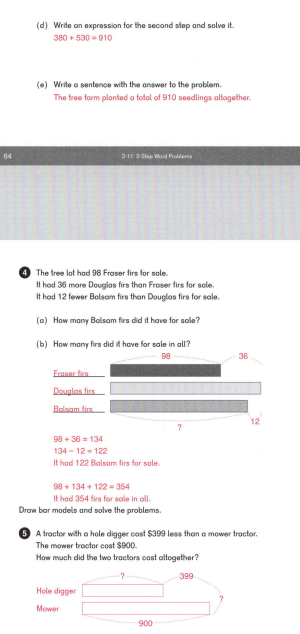

Fraser firs ___ 98 ___ 36
Douglas firs ___
Balsam firs ___ ? ___ 12

98 + 36 = 134
134 − 12 = 122
It had 122 Balsam firs for sale.

98 + 134 + 122 = 354
It had 354 firs for sale in all.

Draw bar models and solve the problems.

5 A tractor with a hole digger cost $399 less than a mower tractor.
The mower tractor cost $900.
How much did the two tractors cost altogether?

Hole digger ___ ? ___ 399
Mower ___ ?
900

$900 − $399 = $501
$501 + $900 = $1,401
The two tractors cost $1,401 altogether.

6 At the time for harvesting the trees, out of 890 trees,
650 were more than 5 ft tall.
The rest were less than 5 ft tall.
How many more trees were taller than 5 ft than were shorter than 5 ft?

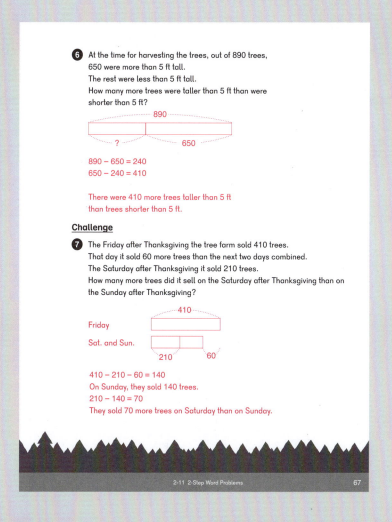

890
? | 650

890 − 650 = 240
650 − 240 = 410

There were 410 more trees taller than 5 ft
than trees shorter than 5 ft.

Challenge

7 The Friday after Thanksgiving the tree farm sold 410 trees.
That day it sold 60 more trees than the next two days combined.
The Saturday after Thanksgiving it sold 210 trees.
How many more trees did it sell on the Saturday after Thanksgiving than on the Sunday after Thanksgiving?

Friday ___ 410
Sat. and Sun. ___
210 | 60

410 − 210 − 60 = 140
On Sunday, they sold 140 trees.
210 − 140 = 70
They sold 70 more trees on Saturday than on Sunday.

Teacher's Guide 3A Chapter 2

Exercise 12

Check

1 Jett scored 640 points in a game.
The game had 3 levels.
He scored the same score, 270, for the first and third level of the game.
How much did he score for the second level of the game?

640

First	Second	Third
270	?	270

640 − 270 − 270 = 100
He scored 100 points on the second level.

2 At a fair, 410 balloons were sold on Saturday.
90 more balloons were sold on Sunday than Saturday.
How many balloons were sold that weekend?

410   90

Saturday [ ]
Sunday [ ]   ?

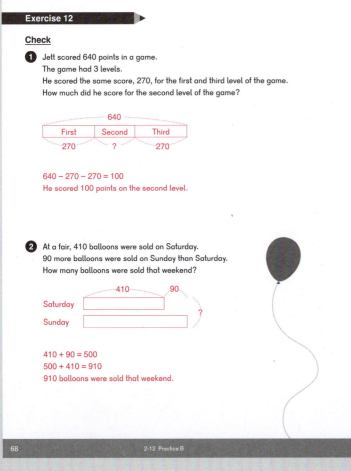

410 + 90 = 500
500 + 410 = 910
910 balloons were sold that weekend.

3 Ada saved $100 over a period of 2 months.
The first month she saved $38.
How much more money did she save the second month than the first month?

38   ?
[ ]   100
[   ]

$100 − $38 = $62
$62 − $38 = $24
She saved $24 more the second month
than the first month.

4 A store has 1,000 light bulbs.
130 of them are incandescent bulbs.
430 of them are fluorescent bulbs.
The rest are LED bulbs.

(a) How many of the bulbs are LED bulbs?

(b) How many fewer LED bulbs are there than the other two types of bulbs?

1,000

Incandescent	Fluorescent	LED
130	430	?

1,000 − 130 − 430 = 440
440 of the bulbs are LED bulbs.

130 + 430 = 560
560 − 440 = 120
There are 120 fewer LED bulbs than the other two kinds.

5 Santiago has 360 coins in his coin collection.
198 of them are foreign coins and the rest are domestic coins.
How many fewer domestic coins does he have than foreign coins?

198

Foreign [ ]
360
Domestic [ ]
?

360 − 198 = 162
198 − 162 = 36
He has 36 fewer domestic coins
than foreign coins.

6 For a book sale, the books were sorted into non-fiction and fiction.
The non-fiction books were further sorted into biographies and other.
There were 500 fiction books and 200 biographies.
There were 50 fewer non-fiction than fiction books.

(a) How many non-fiction books that were not biographies were there?

(b) How many books were there in all?

500

Fiction books [ ]

Non-fiction | biographies | other |
200   ?   50

500 − 200 − 50 = 250
There were 250 non-fiction books that are not biographies.

500 + 200 + 250 = 950
There were 950 books in all.

7 At the book sale, 230 books were sold on Saturday.
70 fewer books were sold on Friday than on Saturday.
50 more books were sold on Sunday than on Friday.

(a) How many more books were sold on Saturday than on Sunday?

(b) How books were sold on Sunday?

(c) How many books were sold in all for the three days?

70

Friday [ ]
Saturday [230]
Sunday [ ]
50   ?

70 − 50 = 20
20 more books were sold on Saturday than on Sunday.

230 − 20 = 210
210 books were sold on Sunday.

Friday: 230 − 70 = 160
160 + 230 + 210 = 600
600 books were sold in all.

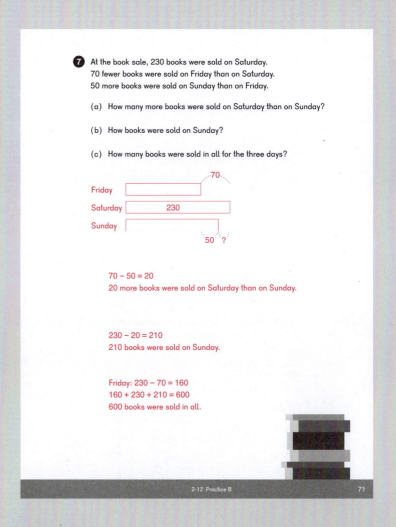

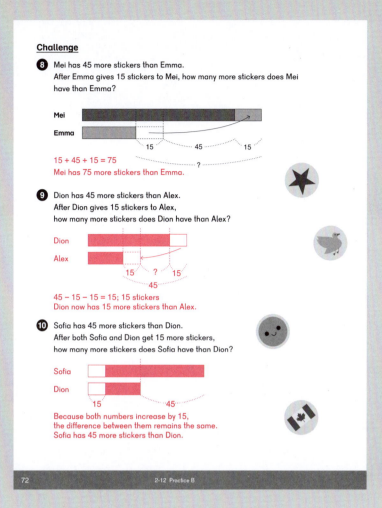

Challenge

8 Mei has 45 more stickers than Emma.
After Emma gives 15 stickers to Mei, how many more stickers does Mei have than Emma?

Mei

Emma

15   45   15

$15 + 45 + 15 = 75$
Mei has 75 more stickers than Emma.

9 Dion has 45 more stickers than Alex.
After Dion gives 15 stickers to Alex,
how many more stickers does Dion have than Alex?

Dion

Alex

15   ?   15
45

$45 - 15 - 15 = 15$; 15 stickers
Dion now has 15 more stickers than Alex.

10 Sofia has 45 more stickers than Dion.
After both Sofia and Dion get 15 more stickers,
how many more stickers does Sofia have than Dion?

Sofia

Dion

15   45

Because both numbers increase by 15,
the difference between them remains the same.
Sofia has 45 more stickers than Dion.

72   2-12 Practice B

Notes

Suggested number of class periods: 7–8

	Lesson	Page	Resources		Objectives
	Chapter Opener	p. 99	TB:	p. 73	Investigate addition and subtraction of four-digit numbers.
1	Addition with Regrouping	p. 100	TB: WB:	p. 74 p. 73	Add two numbers within 10,000 where regrouping occurs in more than one place.
2	Subtraction with Regrouping — Part 1	p. 103	TB: WB:	p. 78 p. 76	Subtract two numbers within 10,000 where regrouping occurs in more than one place.
3	Subtraction with Regrouping — Part 2	p. 106	TB: WB:	p. 83 p. 79	Subtract two numbers within 10,000 where regrouping occurs across zeros.
4	Estimating Sums and Differences — Part 1	p. 109	TB: WB:	p. 88 p. 83	Use rounding to estimate sums and differences of numbers up to three digits.
5	Estimating Sums and Differences — Part 2	p. 111	TB: WB:	p. 90 p. 85	Use rounding to estimate sums and differences of numbers up to four digits.
6	Word Problems	p. 113	TB: WB:	p. 92 p. 87	Solve two-step word problems involving addition and subtraction to 10,000.
7	Practice	p. 115	TB: WB:	p. 95 p. 91	Practice concepts and skills from the chapter.
	Workbook Solutions	p. 117			

In **Dimensions Math® 2A**, students learned how to add and subtract three-digit numbers using the standard algorithm.

In this chapter, students will:

* Extend their knowledge of the addition and subtraction algorithm.

* Use rounding to find estimated answers to addition or subtraction problems.

* Apply these skills in multi-step word problems.

The standard algorithm is important because of its simplicity. It requires only addition and subtraction within 20.

Students will continue to use place-value discs to enrich their conceptual understanding of the algorithm as they gain procedural fluency. Students should use the place-value discs when first solving the **Think** problems in Lessons 1 through 7.

A general procedure for demonstrating the addition algorithm with place-value discs for four-digit numbers when the sum is less than 10,000 is shown here.

Addition Algorithm

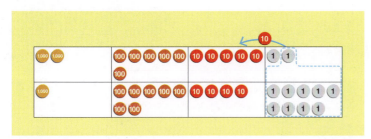

9 ones + 2 ones = 11 ones

Regroup 11 ones as 1 ten and 1 one.

$$\begin{array}{r} 2,652 \\ +1,749 \\ \hline \end{array}$$

Write the digit 1 at the top of the tens column denoting the regrouped 10 ones, and the digit 1 below the line in the ones column.

$$\begin{array}{r} 1 \\ 2,652 \\ +1,749 \\ \hline 1 \end{array}$$

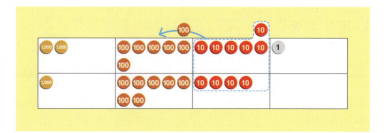

Add the tens, including any regrouped tens:

5 tens + 4 tens + 1 ten = 10 tens

Regroup 10 tens as 1 hundred and 0 tens.

Write the digit 1 at the top of the hundreds column denoting the regrouped 10 tens, and the digit 0 below the line in the tens column.

$$\begin{array}{r} 11 \\ 2,652 \\ +1,749 \\ \hline 01 \end{array}$$

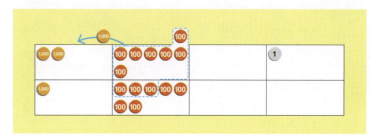

Add the hundreds, including any regrouped hundreds:

6 hundreds + 7 hundreds + 1 hundred = 14 hundreds

Regroup 10 hundreds as 1 thousand and 4 hundreds.

Write the digit 1 at the top of the thousands column denoting the regrouped 10 hundreds, and 4 below the line in the hundreds column.

$$\begin{array}{r} 111 \\ 2,652 \\ +1,749 \\ \hline 401 \end{array}$$

Add the thousands, including any regrouped thousands:

2 thousands + 1 thousand + 1 thousand = 4 thousands.

Write the digit 4 below the line in the hundreds column.

2,652 + 1,749 = 4,401

Subtraction Algorithm

As with addition, students will also use place-value discs while learning the subtraction algorithm.

When subtracting, students may need to regroup from a greater place value and then subtract. This can involve many place-value discs. If students have difficulty keeping track of how many discs are subtracted, or removed from the chart, they can use place-value cards for 1,000, 100, 10, and 1 and put the subtracted discs on them.

This strategy allows students to retain the removed or "subtracted" discs so that they can be added back to the difference to check their work.

Begin with the whole represented on a place-value organizer.

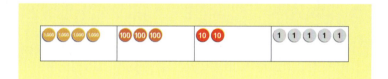

Subtract the ones:

There are not enough ones to subtract 8 ones.

$$\begin{array}{r} 4,325 \\ -2,468 \\ \hline \end{array}$$

Regroup 1 ten as 10 ones.

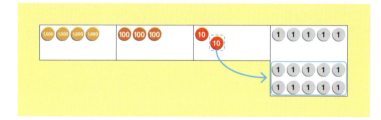

Cross off the digit 2 in the tens place and write a 1 above it.

Cross off the digit 5 in the ones place and write 15 above it.

$$\begin{array}{r} {\scriptstyle 1\,15} \\ 4,3\cancel{2}\cancel{5} \\ -2,468 \\ \hline 7 \end{array}$$

There are now 15 ones.

15 ones − 8 ones = 7 ones.

Write the difference of 7 ones below the line in the ones column.

Subtract the tens:

There are not enough tens to subtract 6 tens.

Regroup 1 hundred as 10 tens.

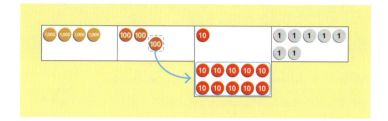

Cross off the digit 3 in the hundreds place and write a 2 above it.

Write the 11 in the tens place to represent the 11 tens.

$$\begin{array}{r} {\scriptstyle 2\,11\,15} \\ 4,\cancel{3}\cancel{2}\cancel{5} \\ -2,468 \\ \hline 57 \end{array}$$

11 tens − 6 tens = 5 tens. Write difference of 5 tens below the line in the tens place.

Subtract the hundreds:

There are not enough hundreds to subtract 4 hundreds.

Regroup 1 thousand as 10 hundreds.

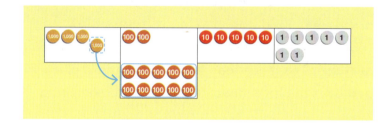

Cross off the digit 4 in the thousands place and write a 3 above it.

Write 12 in the hundreds place to represent the 12 hundreds.

$$\begin{array}{r} {\scriptstyle 3\,12\,11\,15} \\ \cancel{4}\cancel{,3}\cancel{2}\cancel{5} \\ -2,468 \\ \hline 857 \end{array}$$

12 hundreds − 4 hundreds = 8 hundreds

Write the difference of 8 hundreds below the line in the hundreds place.

Subtract the thousands:

3 thousand − 2 thousand = 1 thousand

Write the difference of 1 thousand below the line in the thousands place.

$$\begin{array}{r} {\scriptstyle 3\ 12\ 11\ 15} \\ \cancel{4,3\,2\,5} \\ -\ 2,4\,6\,8 \\ \hline 1,8\,5\,7 \end{array}$$

Students should have one 1,000-disc, eight 100-discs, five 10-discs, and seven 1-discs on their place-value organizers.

They can check their work by adding the two parts together:

$$\begin{array}{r} 4,3\,2\,5 \ \ \text{whole} \\ -\ 2,4\,6\,8 \ \ \text{part} \\ \hline 1,8\,5\,7 \ \ \text{part} \end{array} \qquad \begin{array}{r} 1,8\,5\,7 \ \ \text{part} \\ +\ 2,4\,6\,8 \ \ \text{part} \\ \hline 4,3\,2\,5 \ \ \text{whole} \end{array}$$

When explaining the addition and subtraction processes, avoid using the terms "carrying" or "borrowing," or phrases such as, "more on the floor, go next door," as it can lead to confusion about place value and does not help students understand regrouping.

Ensure the terms regrouping, renaming, trading, or exchanging are used, and emphasize the place being regrouped to or from.

In both the addition and subtraction algorithms, students should recognize the line is acting as an equal sign, where the expression on one side has the same value as the expression on the other side.

$$4{,}325 - 2{,}468 = 1{,}857 \qquad \begin{array}{r} 4,3\,2\,5 \\ -\ 2,4\,6\,8 \\ \hline 1,8\,5\,7 \end{array}$$

Note on managing manipulatives

Typically, students can solve **Do** problems pictorially.

Due to the amount of manipulatives used in this chapter, some suggestions on structuring the lesson are included below:

- Pose the **Think** problem and allow adequate time for students to work out solutions. After they solve the problem, have students share and discuss their methods. The textbook does not need to be open at this time.

- Model the **Learn** procedure, and have students work the problem with place-value discs while it's being modeled.

- Open the textbook and discuss the **Learn** section, making sure students understand how to relate the steps they have done with place-value discs to the written algorithm.

- Struggling students may work the **Do** problems using the place-value discs **first**, then compare them to the textbooks.

- Have students work in pairs with the place-value discs, whiteboards, and textbooks as each student or pair of students will need twenty 1-discs, twenty 10-discs, twenty 100-discs, and nine 1,000-discs.

- As students gain confidence, let them copy any remaining problems from the textbook into notebooks or whiteboards and work them without place-value discs. Students could also draw place-value discs as an interim step.

The goal is to have the students work the problems without place-value discs.

Estimation

In Chapter 1, students learned to round to the nearest 10, 100, and 1,000. They will use this skill to round numbers in order to estimate an answer.

In real-life situations, such as shopping, we may want to round numbers in such a way that we know we are overestimating or underestimating, depending on the circumstances.

Estimated answers will vary depending on how the numbers are rounded, but numbers should be rounded in a way that the estimate is reasonably close to the actual answer.

In this chapter, students will be shown two different rounding strategies. They can round each number to the greatest place for both numbers. For example, rounding 4,272 to 4,000 and 696 to 700. Students can also round both numbers to the greatest place for the lesser number. For example, rounding 696 to 700, then 4,272 to 4,300. Students might also come up with some alternative ways to round the numbers depending on their proficiency with mental math strategies.

The emphasis should be on students using methods that are natural and easy for them to estimate mentally, not on a formal procedure for estimation.

Materials

- 10-sided die
- 3-minute timer
- Dry erase markers
- Place-value discs
- Playing cards
- Whiteboards

Blackline Masters

- Add 'em Up — 4-Digit
- Add 'em Up Number Cards
- Inch Graph Paper
- Number Cards
- Place-value Organizer

Activities

Games and activities included in this chapter are designed to provide practice with computing and estimation for addition and subtraction of four-digit numbers. They can be used after students complete the **Do** questions, or any time additional practice is needed.

Chapter Opener

Objective

- Investigate addition and subtraction of four-digit numbers.

Have students choose two instruments from the illustration on textbook page 73, and discuss whether they have already learned to add or subtract those two numbers. Then, ask students to round the two numbers to find about how much money they would need to buy the two instruments. Ask:

- Did you round up to make sure you have enough money?
- Are your two instruments going to cost more or less than $500?
- How can you find out?

Students can discuss which instruments they chose and how they rounded the numbers. They can also try to find the actual costs using known strategies. Examples:

- I picked the pink guitar and the cymbals so I could add them together using mental math: $349 + $294 = $343 + $300 = $643.
- I rounded the cost of the pink guitar and cymbals to the nearest hundred. $300 + $300 = $600. But the answer seemed wrong: $300 − $300 = 0. So I tried again and rounded to a ten: $350 − $300 = $50.
- I rounded the cost of the two guitars to the nearest hundreds: $1,900 + $300 = $2,200 and $1,900 − $300 = $1,600.

Activities

▲ First to 1,000

Materials: Place-value Organizer (BLM), 4 sets of Number Cards (BLM) 0 to 9, place-value discs

Students take turns drawing two Number Cards (BLM). On each turn, players make a two-digit number with the cards, then add that number of place-value

Chapter 3

Addition and Subtraction — Part 2

$349 $3,685 $294 $100 $639 $849 $2,899 $1,189

Pick two instruments. Do you know how to find the sum of their costs and the difference between their costs?

Round the numbers, then add or subtract to find the approximate sum and difference.

Which numbers do I already know how to add or subtract?

Which place should I round each number to?

73

discs to their Place-value Organizer (BLM).

Any time students accumulate more than nine of any value discs, they exchange ten of them for one of the next greater value.

The winner is the first player to exchange ten 100-discs for a 1,000-disc.

▲ First to 0

Materials: Place-value Organizer (BLM), 4 sets of Number Cards (BLM) 0 to 9, place-value discs

Play is similar to **First to 1,000**. Begin the game with one 1,000-disc and subtract the value of the two-digit number formed with the Number Cards (BLM) on each turn. The winner is the first player to be out of discs.

Lesson 1 Addition with Regrouping

Objective

- Add two numbers within 10,000 where regrouping occurs in more than one place.

Lesson Materials

- Place-value discs
- Place-value Organizer (BLM)
- 1-inch Graph Paper (BLM)

Think

Provide each student with place-value discs and a Place-value Organizer (BLM) and allow them adequate time to work on a solution to the **Think** problem.

Discuss students' strategies for solving the problem. They should find that they can follow a process similar to what they learned in adding three-digit numbers, simply extending that process to one more place.

Questions to ask students:

- What do we do when we have more than 9 discs in a column?
- In which column do we put the regrouped 1,000-disc?

Working the proper sequence of steps with place-value discs on a Place-value Organizer (BLM) will allow them to understand the steps used in the written algorithm.

Learn

Model the procedure shown in **Learn** for students. For detailed instructions, please consult the chapter notes. It is best to demonstrate or model first, with kids working along with the teacher.

Students do not need to write the numerals of the vertical algorithm until after they have worked the problem with the place-value discs at least once.

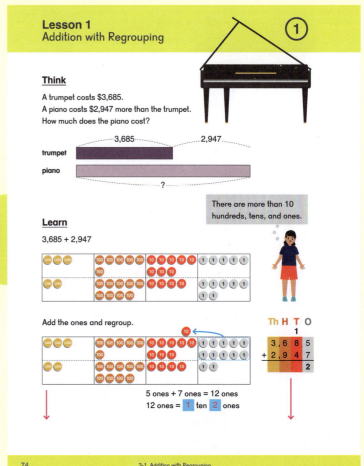

After the students have completed the problem with place-value discs, have them compare the methods they came up with in **Think** with the method shown in the textbook.

Note that:

> 3 thousands   6 hundreds   8 tens   5 ones
> + 5 thousands   9 hundreds   4 tens   7 ones
> ──
> 5 thousands 15 hundreds 12 tens 12 ones
> or 6,632.

Questions to ask students:

- Where do we show regrouping when we work out the problem on paper?
- What does the 1 above the 3 in the thousands column stand for? Where did you see this 1 in the place-value discs?

Point out to students where the regrouped 1,000-disc is recorded in the written equation.

Do

Students can work these problems with the place-value discs first. Note the progression in ❶ is designed to help students see how to regroup in each place.

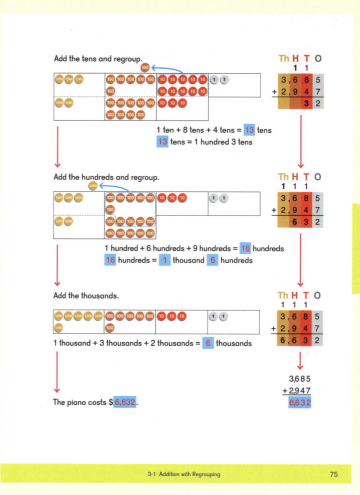

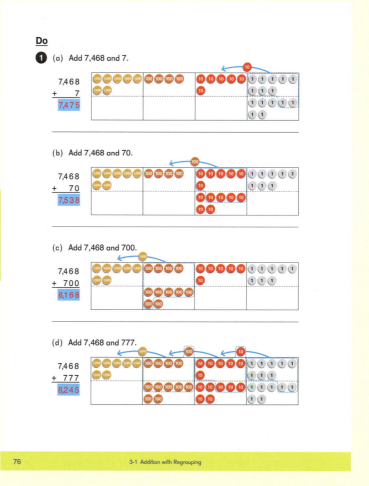

2 — **3** Students should rewrite these problems vertically.

2 (a) and (b) Check that students are aligning the digits correctly. Inch Graph Paper (BLM) or notebook paper turned sideways can be used to help students line up numbers in the correct place value columns.

			2	3	5	6		
		+			8	6		

4 Ask students:

- In what places do you think there will be regrouping? Why?
- In what place do you want to begin solving the problem? Why?
- What does that tell us about the digit that will go in the thousands place?

5 Ask students where they would label the model with a question mark.

Activity

▲ **7,001 Up**

Materials: Number Cards (BLM) 1 to 9 or playing cards

Each player starts with 1,000 as a start number.

On their turn, a player draws three Number Cards (BLM) and makes a three-digit number with the cards. She adds that number to her start number to create a new start number.

The winner is the first player whose running total exceeds the number 7,001.

Exercise 1 • page 73

2 Find the value.

(a) 2,356 + 86   2,442   (b) 6,962 + 93   7,055

(c) 8,480 + 694   9,174   (d) 7,028 + 909 + 60   7,997

(e) 5,272 + 3,764   9,036   (f) 3,547 + 1,793 + 436   5,776

3 What number is 888 more than 6,666?   7,554

4 What are the missing digits?

(a)
```
  4 , 4 2 3
+     7 2 8
  5 , 1 5 1
```

(b)
```
  6 , 3 9 5
+ 2 , 6 0 5
  9 , 0 0 0
```

5 A school bought new musical equipment for the jazz band. The amplifier cost $1,189, the conga drum cost $639, and the chimes cost $294.
How much did all three instruments cost? $1,189 + $639 + $294 = $2,122

Exercise 1 • page 73

3-1 Addition with Regrouping   77

Lesson 2 Subtraction with Regrouping — Part 1

Objective

- Subtract two numbers within 10,000 where regrouping occurs in more than one place.

Lesson Materials

- Place-value discs
- Place-value Organizer (BLM)

Think

Provide each student with place-value discs and a Place-value Organizer (BLM) and allow them adequate time to work on a solution to the **Think** problem.

Ask students, "What can you do when you have to regroup more than once?" (Trade one disc for ten discs of the next lesser value. Note the regrouping above the correct place.)

Discuss student strategies for solving the problem. They should notice that subtracting four-digit numbers follows a similar process to subtracting three-digit numbers.

Learn

Model the procedure shown in **Learn** for students. Students should all have place-value discs and a Place-value Organizer (BLM) on their desks. Have students work along as the steps are being modeled.

After the students have completed the problem with place-value discs, have them compare the methods they came up with in **Think** with the method shown in the textbook.

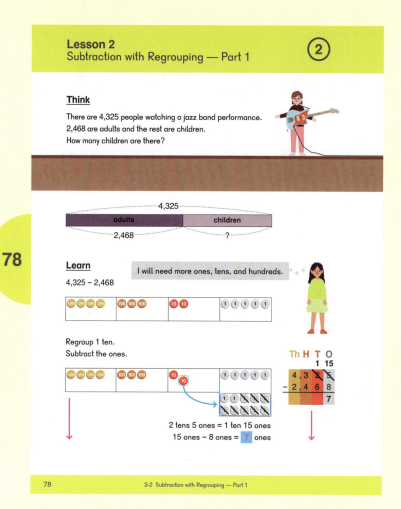

78

Note that:

> 3 thousands 12 hundreds 11 tens 15 ones
> − 2 thousands   4 hundreds   6 tens   8 ones
> _____
>   1 thousand   8 hundreds   5 tens   7 ones
> or 1,857.

Questions to ask students:

- What do we do when we do not have enough place-value discs in a column to subtract from?
- How do we record the regrouped place-value discs in the written equation?
- How can we check our work?

Do

Students should use the pictures of the place-value discs to think about the problems. Allow struggling students to use place-value discs if needed.

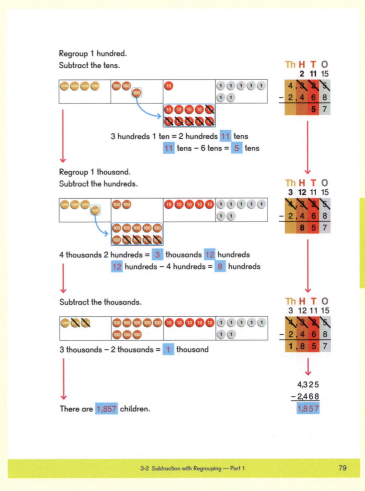

Regroup 1 hundred.
Subtract the tens.

3 hundreds 1 ten = 2 hundreds 11 tens
11 tens − 6 tens = 5 tens

Regroup 1 thousand.
Subtract the hundreds.

4 thousands 2 hundreds = 3 thousands 12 hundreds
12 hundreds − 4 hundreds = 8 hundreds

Subtract the thousands.

3 thousands − 2 thousands = 1 thousand

There are 1,857 children.

```
 4,325
−2,468
 1,857
```

79

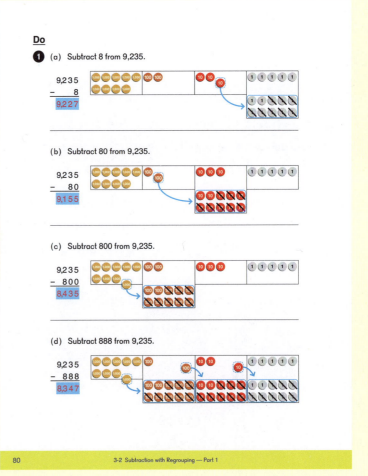

Do

1 (a) Subtract 8 from 9,235.

```
 9,235
−    8
 9,227
```

(b) Subtract 80 from 9,235.

```
 9,235
−   80
 9,155
```

(c) Subtract 800 from 9,235.

```
 9,235
−  800
 8,435
```

(d) Subtract 888 from 9,235.

```
 9,235
−  888
 8,347
```

80

2 Alex reminds students that they can add to check that their answers are correct.

4 Students should rewrite these problems vertically. For (a) and (b), check that students are lining up the numbers in the correct places.

6 Based on part-whole understanding of numbers, we know the whole. To find the missing part, we can subtract the known part from the whole, or 5,530 – ▊ = 168 is the same as 5,530 – 168 = ▊.

8 Ask students where they would label the model with a question mark.

If needed, have them redraw the model on a whiteboard to see the two steps.

Exercise 2 • page 76

---

2 Subtract 2,683 from 6,148.

$$\begin{array}{r} 6,148 \\ -\ 2,683 \\ \hline 3,465 \end{array}$$

$$\begin{array}{r} 6,148 \\ -\ 2,683 \\ \hline 3,465 \end{array} \qquad \begin{array}{r} 3,465 \\ +\ 2,683 \\ \hline 6,148 \end{array}$$

Add to check the answer.

3 Find the value of 8,627 – 3,719.

$$\begin{array}{r} 8,627 \\ -\ 3,719 \\ \hline 4,908 \end{array}$$

4 Find the value.

(a) 2,347 – 168   2,179

(b) 1,419 – 729   690

(c) 4,159 – 1,361   2,798

(d) 8,260 – 3,479   4,781

(e) 4,107 – 1,046   3,061

(f) 8,096 – 3,575   4,521

---

5 What are the missing digits?

(a)
$$\begin{array}{r} 8,\boxed{5}\,4\,9 \\ -\quad\ 8\,2\,3 \\ \hline \boxed{7},7\,2\,\boxed{6} \end{array}$$

(b)
$$\begin{array}{r} \boxed{7},9\,3\,\boxed{5} \\ -\ 3,3\,\boxed{4}\,2 \\ \hline 4,5\,9\,3 \end{array}$$

6 What are the missing numbers?

(a) 5,530 – $\boxed{5,362}$ = 168

(b) 5,092 – $\boxed{2,929}$ = 2,163

7 Form the greatest 4-digit number using each of the following digits once. Form the least 4-digit number using each of the following digits once. Then, find the difference between the two numbers.

| 4 | 8 | 1 | 3 |

Greatest: 8,431
Least: 1,348
Difference: 7,083

8 The difference between two numbers is 1,498. The greater number is 2,185.

2,185

(a) What is the other number?
2,185 – 1,498 = 687, 687

(b) What is the sum of the two numbers?
2,185 + 687 = 2,872, 2,872

1,498

Exercise 2 • page 76

Lesson 3 Subtraction with Regrouping — Part 2

Objective

- Subtract two numbers within 10,000 where regrouping occurs across zeros.

Lesson Materials

- Place-value discs
- Place-value Organizer (BLM)

Think

Provide each student with place-value discs and a Place-value Organizer (BLM) and allow them adequate time to work on the **Think** problem.

Ask students:

- How is this problem different from the ones you solved in the previous lesson?
- How is it the same? (We still subtract starting at the ones place.)
- What can you do when you have to regroup more than once?

Students should note that there are two places with no value. They will need to:

- Regroup from the thousands place to the hundreds place first, then
- Regroup from the hundreds place to the tens place, then
- Regroup from the tens place to the ones place to have enough ones to subtract 9 ones.

Discuss student strategies for solving the problem.

Learn

Model the procedure shown in **Learn** for students. Students should all have place-value discs and a Place-value Organizer (BLM) on their desks. Have students work along as the steps are being modeled.

Discuss Emma's comment. Ask students what they can do if there are no tens.

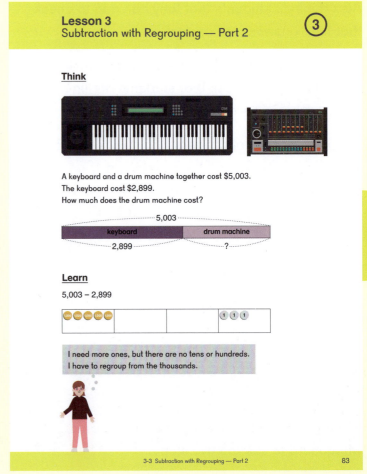

After the students have completed the problem with place-value discs, have them compare the methods they came up with in **Think** with the method shown in the textbook.

Note that:

4 thousands	9 hundreds	9 tens	13 ones
− 2 thousands	8 hundreds	9 tens	9 ones
2 thousands	1 hundred	0 tens	4 ones

or 2,104.

Questions to ask students:

- What do we do when we do not have enough place-value discs in a column to subtract from?
- How do we record the regrouped place-value discs in a written equation?
- Why is there a 10 and a 9 above both the hundreds and tens places?
- How can we check our work?

Have students use addition to check their work.

Do

Students can work these problems with place-value discs first as needed.

2 Alex sees that he can express the number 5,006 as 499 tens and 16.

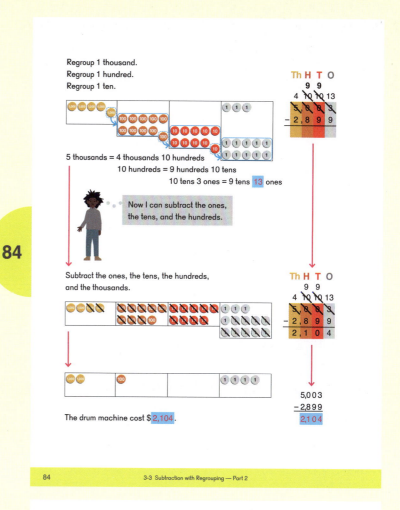

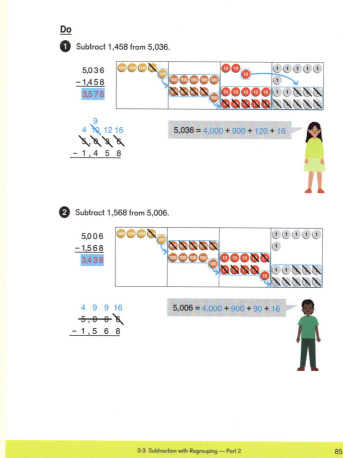

4 Alex is showing how he could solve this problem with a mental math strategy.

6 Students should work these problems with the vertical algorithm and check their calculations with a mental math strategy.

▲ Activity

7,001 Out

Materials: Number Cards (BLM) 1 to 9 or playing cards

Each player starts with 7,001 as a start number.

On his turn, a player draws three Number Cards (BLM) and makes a three-digit number with them. He subtracts that number from the start number to create a new start number.

The winner is the first player whose running total falls below 100.

Exercise 3 • page 79

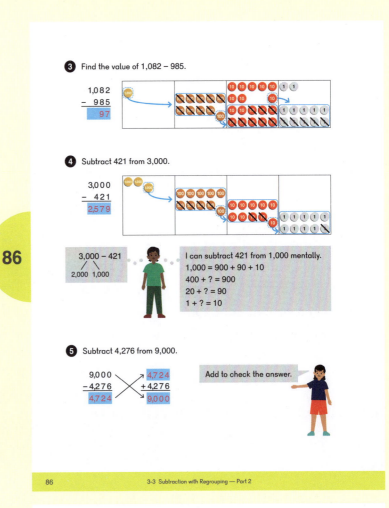

86

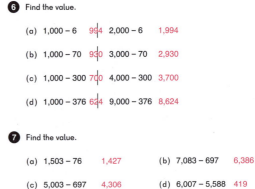

6 Find the value.

(a) 1,000 − 6   994   2,000 − 6   1,994

(b) 1,000 − 70   930   3,000 − 70   2,930

(c) 1,000 − 300   700   4,000 − 300   3,700

(d) 1,000 − 376   624   9,000 − 376   8,624

7 Find the value.

(a) 1,503 − 76   1,427   (b) 7,083 − 697   6,386

(c) 5,003 − 697   4,306   (d) 6,007 − 5,588   419

(e) 5,000 − 3,333   1,667   (f) 10,000 − 4,714   5,286

87

8 What are the missing digits?

(a)   3,8 0 2
    −   5 2 7
      3,2 7 5

(b)   6,0 2 9
    − 3,3 3 3
      2,6 9 6

Exercise 3 • page 79

Lesson 4 Estimating Sums and Differences — Part 1

Objective

- Use rounding to estimate sums and differences of numbers up to three digits.

Think

Pose the **Think** problem and allow students adequate time to find an estimate. Have them discuss why they rounded as they did and whether they think their estimate is higher or lower than the actual value.

Examples:

- I thought of $349 as $350 and added $60, so I think my estimate is a little lower than the actual cost.
- I rounded both numbers to the nearest hundred and estimate the guitar and the guitar case will probably cost less than $450.
- I rounded to $350 and $70 and I think my answer is pretty close.

Ask students how Sofia and Dion got their estimates.

Learn

Dion rounded the number $349 to the nearest hundred and $67 to the nearest ten. Sofia rounded both numbers to the nearest ten.

Have students discuss how close their estimates are to the exact answer, and which method works best for them.

Estimates can be used to check if an answer is reasonable.

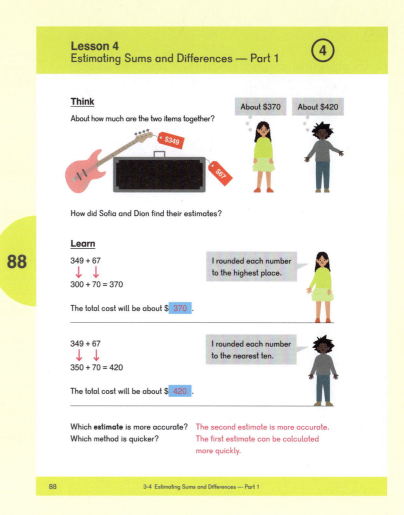

88

Do

3 Have students estimate and then find the exact answer for some of these problems to compare how close their estimates were to the exact answers.

Students can round the numbers differently in order to get an estimate.

Exact solutions:

(a) 598   (b) 832   (c) 632
(d) 618   (e) 299   (f) 115
(g) 486   (h) 298   (i) 862

Exercise 4 • page 83

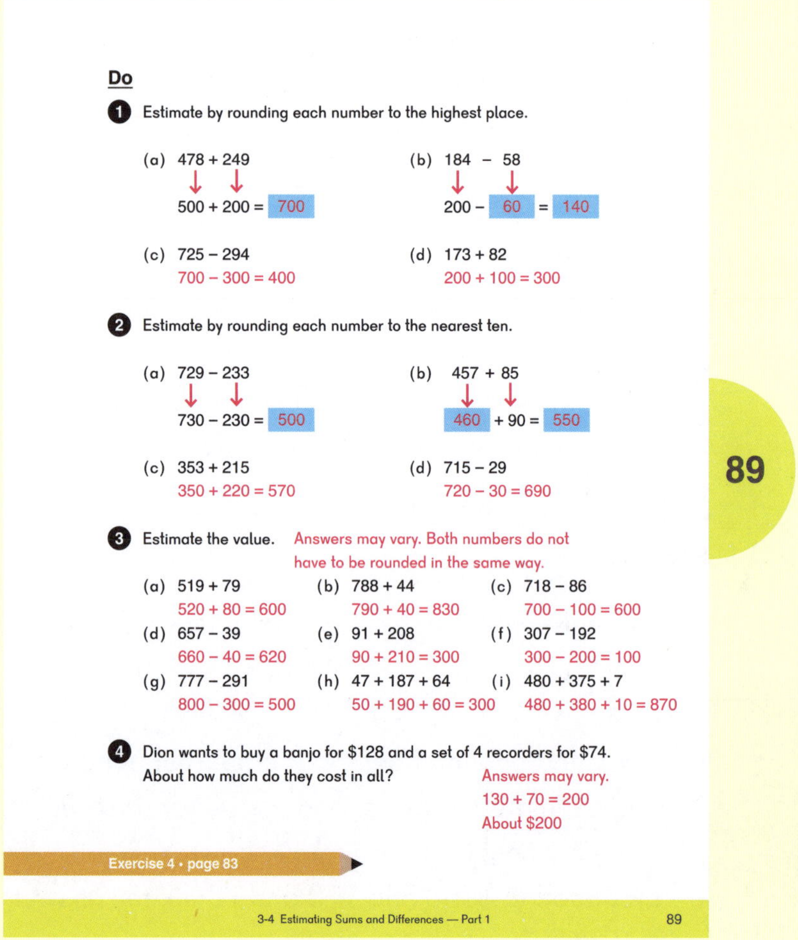

Do

1 Estimate by rounding each number to the highest place.

(a) 478 + 249
  ↓   ↓
500 + 200 = 700

(b) 184 − 58
  ↓   ↓
200 − 60 = 140

(c) 725 − 294
700 − 300 = 400

(d) 173 + 82
200 + 100 = 300

2 Estimate by rounding each number to the nearest ten.

(a) 729 − 233
  ↓   ↓
730 − 230 = 500

(b) 457 + 85
  ↓   ↓
460 + 90 = 550

(c) 353 + 215
350 + 220 = 570

(d) 715 − 29
720 − 30 = 690

3 Estimate the value. Answers may vary. Both numbers do not have to be rounded in the same way.

(a) 519 + 79
520 + 80 = 600

(b) 788 + 44
790 + 40 = 830

(c) 718 − 86
700 − 100 = 600

(d) 657 − 39
660 − 40 = 620

(e) 91 + 208
90 + 210 = 300

(f) 307 − 192
300 − 200 = 100

(g) 777 − 291
800 − 300 = 500

(h) 47 + 187 + 64
50 + 190 + 60 = 300

(i) 480 + 375 + 7
480 + 380 + 10 = 870

4 Dion wants to buy a banjo for $128 and a set of 4 recorders for $74. About how much do they cost in all?

Answers may vary.
130 + 70 = 200
About $200

Exercise 4 • page 83

89

Lesson 5 Estimating Sums and Differences — Part 2

Objective

- Use rounding to estimate sums and differences of numbers up to four digits.

Think

Pose the **Think** problem and allow students adequate time to find an estimate. Have them discuss why they rounded as they did and whether they think their estimates are higher or lower than the actual value.

Examples:

- I thought of $4,275 as $4,300 and $696 as $700. $4,300 − $700 = $3,600. I rounded both numbers up, so I think my estimate is a little more than the actual cost.
- I rounded both numbers to the nearest thousand and subtracted $4,000 − $1,000 so I estimate the Solar banjo will cost about $3,000 more than the Goldenrod banjo.
- I rounded $4,275 to the nearest thousand and $696 to the nearest hundred. $4,000 − 700 = $3,300 and I think my answer is pretty close.

Learn

Have students note how close their estimates are to the actual answers, and consider situations when it might be reasonable to use one estimate or the other. For example:

- A shopper might overestimate the cost of a purchase to ensure he has enough money to spend.
- An employee might underestimate her salary and be happy when she is paid more than expected.

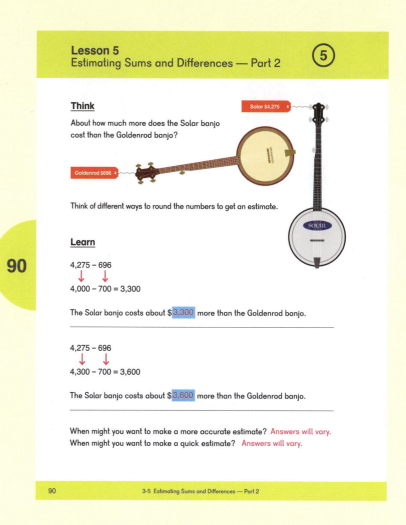

90

Do

3 Have students estimate and then find the exact answer for some of these problems to compare how close their estimate was to the exact answer.

Exact solutions:

(a) 6,024   (b) 3,465   (c) 6,077

(d) 5,860   (e) 1,335   (f) 6,080

(g) 3,137   (h) 7,916   (i) 2,906

Exercise 5 • page 85 ▶

Do

1 Estimate by rounding each number to the highest place.

(a) 6,542 + 3,109
↓   ↓
7,000 + 3,000 = 10,000

(b) 4,462 − 537
↓   ↓
4,000 − 500 = 3,500

(c) 6,072 − 2,700
6,000 − 3,000 = 3,000

(d) 2,731 + 660
3,000 + 700 = 3,700

2 Estimate by rounding each number to the nearest hundred.

(a) 6,502 − 2,360
↓   ↓
6,500 − 2,400 = 4,100

(b) 4,285 + 468
↓   ↓
4,300 + 500 = 4,800

(c) 3,278 + 3,657
3,300 + 3,700 = 7,000

(d) 6,832 − 730
6,800 − 700 = 6,100

3 Estimate the value.   Answers may vary.

(a) 5,396 + 628
5,400 + 600 = 6,000

(b) 2,488 + 977
2,500 + 1,000 = 3,500

(c) 6,721 − 644
6,700 − 600 = 6,100

(d) 3,560 + 2,300
3,600 + 2,300 = 5,900

(e) 2,634 − 1,299
2,600 − 1,300 = 1,300

(f) 4,753 + 1,327
5,000 + 1,000 = 6,000

(g) 8,059 − 4,922
8,000 − 5,000 = 3,000

(h) 7,876 + 40
7,900 + 40 = 7,940

(i) 1,384 + 975 + 547
1,400 + 1,000 + 500 = 2,900

4 A school raised $2,859 at a fund drive and $735 at a bake sale. About how much money did the school raise?   Answers may vary.
2,900 + 700 = 3,600
About $3,600

Exercise 5 • page 85 ▶

Lesson 6 Word Problems

Objective

- Solve two-step word problems involving addition and subtraction to 10,000.

Think

Pose the **Think** problem and allow students time to work on a solution.

Learn

Estimates can help students while drawing representative models to solve a word problem.

In order to draw the bar for instrument stands shorter than the bar for music stands, students have to make a rough estimate, and will actually determine the first step for solving the problem if they do so.

They may then realize that 700 is half of 1,400 and that there must have been more music stands, since 580 (instrument stands) is less than 700.

Once they estimate 1,400 − 600 = 800 music stands, they can draw comparison models that represent more music stands than instrument stands.

Students can do the computation:
1,400 − 580 = 820 music stands in all.
820 − 580 = 240 more music stands than instrument stands.

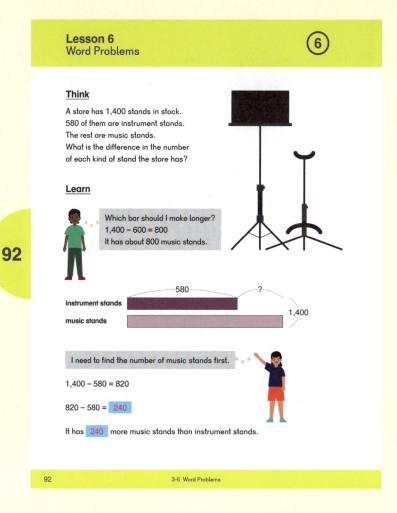

92

Do

1—**3** The models are provided. Students can estimate to check if their answers are reasonable.

5—**6** Draw a representative bar model first, then solve the problem.

Note: While these are comparison problems, students may draw two models depending on how they see the problems.

For example, in **5**, students might read that the baby and mother elephant together weigh 6,104 lb. The baby weighs 205 lb. They could draw part-whole models to find the weight of the mother:

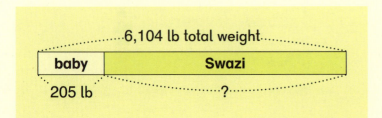

6,104 lb total weight

baby	Swazi

205 lb   ?

A student might then draw a comparison model to find the weight difference between the mother and baby, if needed.

Exercise 6 • page 87

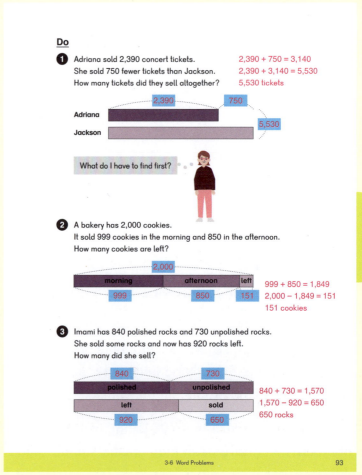

Do

1 Adriana sold 2,390 concert tickets.
She sold 750 fewer tickets than Jackson.
How many tickets did they sell altogether?

2,390 + 750 = 3,140
2,390 + 3,140 = 5,530
5,530 tickets

What do I have to find first?

2 A bakery has 2,000 cookies.
It sold 999 cookies in the morning and 850 in the afternoon.
How many cookies are left?

999 + 850 = 1,849
2,000 − 1,849 = 151
151 cookies

3 Imami has 840 polished rocks and 730 unpolished rocks.
She sold some rocks and now has 920 rocks left.
How many did she sell?

840 + 730 = 1,570
1,570 − 920 = 650
650 rocks

4 Aiyana jogged 975 m.
Brianna jogged 195 m farther than Aiyana.
Clara jogged 400 m less than Briana.

(a) How far did Clara jog?   975 + 195 = 1,170
1,170 − 400 = 770   770 m

(b) How far did the three girls jog in all?   975 + 1,170 + 770 = 2,915
2,915 m

5 Together, Swazi, an African forest elephant, and her baby weigh 6,104 lb.
The baby weighs 205 lb.
How much less does the baby weigh than her mother?   6,104 − 205 = 5,899
5,899 − 205 = 5,694
5,694 lb

6 James, Santiago, and Xavier are collecting bottle caps for a project.
They have collected 3,625 bottle caps so far.
James collected 1,430 bottle caps.
Santiago collected 182 bottle caps fewer than James.

(a) How many bottle caps did James and Santiago collect?

(b) How many fewer bottle caps did Xavier collect than James?

1,430 − 182 = 1,248
1,430 + 1,248 = 2,678
2,678 bottle caps

3,625 − 2,678 = 947
1,430 − 947 = 483
483 bottle caps

Lesson 7 Practice

Objective

- Practice concepts and skills from the chapter.

After students complete the **Practice** in the textbook, have them continue estimating and adding and subtracting four-digit numbers. Students should be fluent with adding and subtracting four-digit numbers before beginning **Chapter 8: Multiplication**.

Activities

▲ Add 'em Up

Materials: 10-sided die, Add 'em Up — 4-Digit (BLM)

Players roll a 10-sided die and put the numbers on Add 'em Up — 4-digit (BLM). Numbers should be placed in the squares as they are rolled to create the parts (addends). Once they have those 8 boxes filled, each player adds to find the sum. The player closest to 8,000 without going over is the winner.

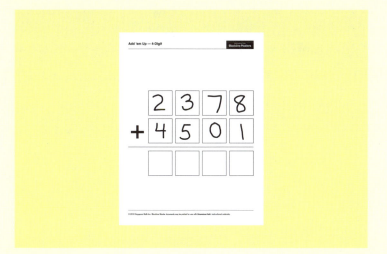

▲ Race to the Sum or Difference

Materials: 4 sets of Add 'em Up Number Cards (BLM), Add 'em Up — 4-Digit (BLM) for each player, 3-minute timer

Number Cards (BLM) are placed facedown in front of the players. Players draw 12 random Number Cards (BLM) each and keep them facedown in front of them.

Lesson 7
Practice
P ⑦

❶ Estimate, then find the value.   Answers may vary.

(a) 1,575 + 783
    1,600 + 800 = 2,400

(b) 1,309 + 3,494
    1,300 + 3,500 = 4,800

(c) 8,976 + 87
    9,000 + 90 = 9,090

(d) 6,428 + 2,539
    6,400 + 2,500 = 8,900

(e) 6,742 − 964
    6,700 − 1,000 = 5,700

(f) 7,204 − 1,207
    7,200 − 1,200 = 6,000

(g) 8,088 − 3,509
    8,000 − 3,500 = 4,500

(h) 5,001 − 94
    5,000 − 100 = 4,900

(i) 4,008 − 1,239
    4,000 − 1,200 = 2,800

(j) 9,000 − 5,855
    9,000 − 6,000 = 3,000

(k) 225 + 750 + 625
    200 + 800 + 600 = 1,600

(l) 1,487 + 265 + 2,277
    1,500 + 300 + 2,000 = 3,800

❷ Janet spent $2,080 one year on guitar lessons and $3,120 on voice lessons.

(a) How much more did the voice lessons cost than the guitar lessons?
    3,120 − 2,080 = 1,040; $1,040

(b) How much did she spend on both types of lessons that year?
    2,080 + 3,120 = 5,200; $5,200

❸ The United States of America was founded on July 4, 1776. On July 1, 1867, Canada was founded. How many years passed between the foundation of the two countries?
1867 − 1776 = 91
91 years

3-7 Practice   95

When both players are ready, they start the timer and flip over their Number Cards (BLM) to reveal their digits. Players work until the timer runs out to create a correct problem.

Note: Players may not be able to use all of their cards. The goal should be to have a completed equation by the end of the 3 minutes.

When time is up players tally their points:

- 1 point for having a correct equation.
- 1 bonus point for using the most cards in their equation.

The winner is the first player to collect 15 points.

Brain Works

★ Cryptarithmetic Puzzles

Each letter stands for a number 0 to 9. Hints are given, if needed. Challenge students with, "Can you write your own?"

Hint: G = 7	Answer
G O O D	7 8 8 1
+ D O G	+ 1 8 7
O R E O	8 0 6 8

Questions for students:

- Notice G and O. How much more is the value of O than the G?
- Does that help us figure out what D is?

Hint: A = 1	Answer
M A K E	3 1 6 5
A	1
+ C A K E	+ 2 1 6 5
E M M A	5 3 3 1

Questions for students:

- What must the E be, if E + E + A equals A?
- How can K + K = M, and A + A = M? Is something being regrouped?

4. The summit of Mount Everest is 8,850 meters above sea level.
The summit of Mount Denali is 2,670 meters lower than the summit of Mount Everest.
How high above sea level is the summit of Mount Denali in meters?
8,850 − 2,670 = 6,180
6,180 m

5. Liam collected 2,324 coins.
He collected 489 coins more than Camilla.
How many coins did they collect altogether?
2,324 − 489 = 1,835
2,324 + 1,835 = 4,159   4,159 coins

6. Yoko saved $950.
She saved $300 more than Malik.
Jasmine saved $200 more than Malik.
How much money did Jasmine save?
950 − 300 = 650
650 + 200 = 850 or: 950 − 100 = 850; $850

7. A store is selling a set of furniture for $1,199.
Sold separately, the couch costs $689, the shelves costs $423, and the side table costs $264.
How much money is saved by buying the set?
689 + 423 + 264 = 1,376
1,376 − 1,199 = 177 $177

8. Heather wants to buy a game console that costs $314.
If she buys it online, it will cost $285.
Estimate the amount of money she will save if she buys it online. About $30

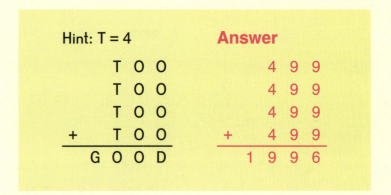

Hint: T = 4	Answer
T O O	4 9 9
T O O	4 9 9
T O O	4 9 9
+ T O O	+ 4 9 9
G O O D	1 9 9 6

Questions for students:

- What value must G be? If 4 Os makes a D in the ones place, how can 4 Os make an O in the tens place?
- Can we regroup more than 1 ten if we add four digits?

Chapter 3 Addition and Subtraction — Part 2

Exercise 1

Basics

1 Find the sum of 3,456 and 5,768.
Start with the ones.

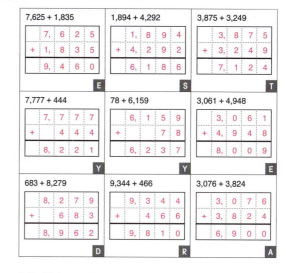

$$\begin{array}{r} 3,4\ 5\ 6 \\ +\ 5,7\ 6\ 8 \\ \hline 9,2\ 2\ 4 \end{array}$$

In which places did you need to regroup?
ones, tens, hundreds

2 Add.

$$\begin{array}{r} 1,7\ 5\ 6 \\ +\ \ \ \ \ 7\ 7 \\ \hline 1,8\ 3\ 3 \end{array} \qquad \begin{array}{r} 8,3\ 3\ 9 \\ +\ \ \ \ 9\ 3\ 5 \\ \hline 9,2\ 7\ 4 \end{array} \qquad \begin{array}{r} 4,6\ 1\ 4 \\ +\ 3,5\ 6\ 7 \\ \hline 8,1\ 8\ 1 \end{array}$$

Practice

3 Add.

7,625 + 1,835	1,894 + 4,292	3,875 + 3,249
7, 6 2 5 + 1, 8 3 5 9, 4 6 0	1, 8 9 4 + 4, 2 9 2 6, 1 8 6	3, 8 7 5 + 3, 2 4 9 7, 1 2 4
E	**S**	**T**
7,777 + 444	78 + 6,159	3,061 + 4,948
7, 7 7 7 + 4 4 4 8, 2 2 1	6, 1 5 9 + 7 8 6, 2 3 7	3, 0 6 1 + 4, 9 4 8 8, 0 0 9
Y	**Y**	**E**
683 + 8,279	9,344 + 466	3,076 + 3,824
8, 2 7 9 + 6 8 3 8, 9 6 2	9, 3 4 4 + 4 6 6 9, 8 1 0	3, 0 7 6 + 3, 8 2 4 6, 9 0 0
D	**R**	**A**

Riddle: What is something you will never see again?
Write the letters that match the answers above to find out.

	Y	E	S	T	E	R	D	A	Y	
8,762	8,221	8,009	6,186	7,124	9,460	9,810	8,962	6,900	6,237	7,224

4 Aki used 1,458 beads for one art project, and
1,905 beads for another art project.
How many beads did she use in all?

$$\begin{array}{r} 1,458 \\ +\ 1,905 \\ \hline 3,363 \end{array} \qquad 3,363 \text{ beads}$$

5 Write the missing digits.

(a)
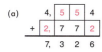
$$\begin{array}{r} 4,\ 5\ 5\ 4 \\ +\ 2,\ 7\ 7\ 2 \\ \hline 7,\ 3\ 2\ 6 \end{array}$$

(b)
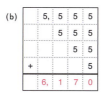
$$\begin{array}{r} 2,\ 5\ 8\ 6 \\ +\ 1,\ 4\ 1\ 4 \\ \hline 4,\ 0\ 0\ 0 \end{array}$$

6 Add.

(a)
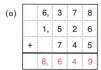
$$\begin{array}{r} 6,\ 3\ 7\ 8 \\ 1,\ 5\ 2\ 6 \\ +\ \ \ \ 7\ 4\ 5 \\ \hline 8,\ 6\ 4\ 9 \end{array}$$

(b)
$$\begin{array}{r} 5,\ 5\ 5\ 5 \\ 5\ 5\ 5 \\ 5\ 5 \\ +\ \ \ \ \ \ \ 5 \\ \hline 6,\ 1\ 7\ 0 \end{array}$$

Challenge

7 Put either 8 or + in each box to make the equation true.
There will be 5 numbers added together.

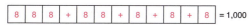

8 8 8 + 8 8 + 8 + 8 + 8 = 1,000

Exercise 2

Basics

1 Subtract 3,397 from 5,483.
Start with the ones.

	5 ,	4	8	3
−	3 ,	3	9	7
	2,	**0**	**8**	**6**

From which places did you need to regroup in order to subtract?

tens, hundreds

2 Subtract.

	9,	7	3	5
−		6	5	7
	9,	**0**	**7**	**8**

	7,	6	3	0
−	1,	8	5	3
	5,	**7**	**7**	**7**

	1,	2	8	9
−		7	8	7
		5	**0**	**2**

Practice

3 Subtract.

4,260 − 335	8,064 − 3,758	7,123 − 1,456
4,260 − 335 = **3,925** (N)	8,064 − 3,758 = **4,306** (I)	7,123 − 1,456 = **5,667** (S)
4,444 − 888	8,230 − 7,765	6,912 − 54
4,444 − 888 = **3,556** (F)	8,230 − 7,765 = **465** (O)	6,912 − 54 = **6,858** (A)
5,307 − 1,226	3,183 − 2,346	7,290 − 5,191
5,307 − 1,226 = **4,081** (R)	3,183 − 2,346 = **837** (C)	7,290 − 5,191 = **2,099** (S)

In what city was the fortune cookie invented?
Write the letters that match the answers above to find out.

S	A	N		F	R	A	N	C	I	S	C	O
5,667	6,858	3,925	5,776	3,556	4,081	6,858	3,925	837	4,306	2,099	837	465

4 Aki had 8,420 beads to use in all.
She has already used 3,363 beads.
How many does she have left?

    8,420
  − 3,363
    5,057   5,057 beads

5 Write the missing digits.

(a)

	4,	1	**7**	3
−		2	6	**7**
	3,	**9**	0	6

(b)

	5,	**0**	7	**0**
−	3,	6	**4**	2
	1,	4	2	8

6 Complete the number pattern.

3,279	2,823	2,367	**1,911**	1,455	**999**

Numbers decrease by 456.

Challenge

7 Find the missing number without calculating 4,388 − 1,672.

4,388 − 1,672 = 5,488 − **2,772**

Exercise 3

Basics

1 Subtract 3,397 from 9,005.
Start with the ones.

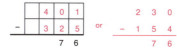

$$
\begin{array}{r}
9,005 \\
-\ 3,397 \\
\hline
5,608
\end{array}
$$

Which place did you start regrouping from first in order to subtract ones?
thousands

2 Subtract.

9, 7 0 5		1, 0 0 2		7, 0 0 9	
− 6 5 7		− 8 5 7		− 4, 0 8 7	
9, 0 4 8		1 4 5		2, 9 2 2	

Practice

3 Subtract.

4,007 − 335	8,004 − 3,758	7,000 − 1,456
4, 0 0 7 − 3 3 5 3, 6 7 2	8, 0 0 4 − 3, 7 5 8 4, 2 4 6	7, 0 0 0 − 1, 4 5 6 5, 5 4 4
1,009 − 888	**8,000 − 3,765**	**6,002 − 56**
1, 0 0 9 − 8 8 8 1 2 1	8, 0 0 0 − 3, 7 6 5 4, 2 3 5	6, 0 0 2 − 5 6 5, 9 4 6
5,017 − 1,996	**3,103 − 2,346**	**7,090 − 5,191**
5, 0 1 7 − 1, 9 9 6 3, 0 2 1	3, 1 0 3 − 2, 3 4 6 7 5 7	7, 0 9 0 − 5, 1 9 1 1, 8 9 9

What is the only number where the number word has the same number of
letters as the number's value?
Color the boxes that match the answers to find out.

3,662	4,235	657	756	5,445	576
899	3,672	4,236	4,672	6,857	1,988
6,544	5,946	221	757	9,865	4,325
3,921	3,021	1,899	121	4,246	4,035
988	4,021	1,757	5,544	3,211	672

4 Aki has 5,057 beads after the last two art projects.
She did a third art project and now has 398 beads left.
How many did she use for her third art project?

$$
\begin{array}{r}
5,057 \\
-\ 398 \\
\hline
4,659
\end{array}
$$
   4,659 beads

5 Complete the number pattern.

5,674	4,786	3,898	3,010	2,122	1,234

Numbers decrease by 888.

6 (a) 6,003 − **1,875** = 4,128

(b) 6,984 + **2,159** = 9,143

(c) **6,983** − 1,897 = 5,086

(d) **3,162** + 2,845 = 6,007

Challenge

7 Put the digits 0, 1, 2, 3, 4, and 5 in the boxes to make the
difference between them 76.

$$
\begin{array}{r}
4\ 0\ 1 \\
-\ 3\ 2\ 5 \\
\hline
7\ 6
\end{array}
\quad \text{or} \quad
\begin{array}{r}
2\ 3\ 0 \\
-\ 1\ 5\ 4 \\
\hline
7\ 6
\end{array}
$$

8 The sum of two numbers is 400.
The digit in the hundreds place of one of the numbers is 1.
The digit in the hundreds place of the other number is 2.

$$2\ \boxed{\ \ }\ \boxed{\ \ } + 1\ \boxed{\ \ }\ \boxed{\ \ } = 4\ 0\ 0$$

(a) What would be the two numbers such that the difference between
them is the greatest?

$$2\ \boxed{9}\ \boxed{9} - 1\ \boxed{0}\ \boxed{1} = 1\ 9\ 8$$

(b) What would be the two numbers such that the difference between
them is the least?

$$2\ \boxed{0}\ \boxed{1} - 1\ \boxed{9}\ \boxed{9} = 2$$

Teacher's Guide 3A Chapter 3

Exercise 4

Basics

1 Estimate the value by rounding each number to the place indicated.

642 – 87

nearest hundred ↓ nearest ten ↓

600 – 90 = 510

642 – 87

nearest ten ↓ nearest ten ↓

640 – 90 = 550

Which estimate is closer to the actual value?

550

2 Estimate the value by rounding each number.

Estimates may vary. Examples shown below.

785 + 84
↓   ↓
800 + 80 = 880

674 – 439
↓   ↓
670 – 440 = 230

895 + 756
↓   ↓
900 + 700 = 1,600

904 – 628
↓   ↓
900 – 600 = 300

Practice

3 Match.
Estimate rather than calculate.

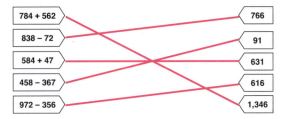

784 + 562 → 616
838 – 72 → 766
584 + 47 → 631
458 – 367 → 91
972 – 356 → 616

784 + 562	766
838 – 72	91
584 + 47	631
458 – 367	616
972 – 356	1,346

4 Write > or < in the ◯.

(a) 807 – 489 (>) 300

(b) 485 + 679 (>) 400 + 700

5 Is 726 – 432 closer to 200 or 300?
300

6 Carlos has 498 foreign coins and 423 domestic coins in his collection.
Grace has 968 foreign coins and 75 domestic coins in her collection.
Who has more coins?
Students may realize they can use estimation to determine that Carlos has less than 1,000 coins and Grace has more than 1,000 coins.

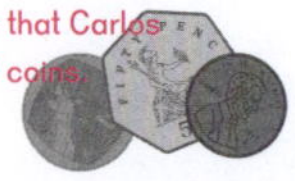

Grace has more coins.

Exercise 5

Basics

1 Estimate the value by rounding each number to the place indicated.

4,223 + 758

nearest thousand ↓ nearest hundred ↓

4,000 + 800 = 4,800

4,223 + 758

nearest hundred ↓ nearest hundred ↓

4,200 + 800 = 5,000

Which estimate is closer to the actual value?

5,000

2 Estimate the value by rounding each number.

Estimates may vary. Examples shown below.

7,412 + 698
↓   ↓
7,000 + 700 = 7,700

9,523 – 4,298
↓   ↓
9,500 – 4,300 = 5,200

3,875 + 3,169
↓   ↓
4,000 + 3,000 = 7,000

6,043 – 78
↓   ↓
6,000 – 80 = 5,920

Practice

3 Match. Estimate rather than calculate.

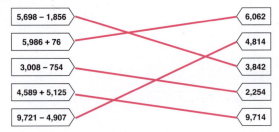

5,698 – 1,856	6,062
5,986 + 76	4,814
3,008 – 754	3,842
4,589 + 5,125	2,254
9,721 – 4,907	9,714

4 Write > or < in the ◯.

(a) 5,894 + 3,125 (>) 9,000

(b) 9,098 – 679 (>) 9,000 – 700

5 Is 4,299 + 3,467 closer to 7,000 or 8,000?
8,000

6 Computer A normally costs $2,413, but is on sale for $450 less.
Computer B normally costs $3,183, but is on sale for $699 less.
Which computer costs less with the sale?
Students may realize they can estimate. Sale cost for Computer A will be a little less than $2,000, and for Computer B will be a little over $2,000.
Computer A costs less with the sale.

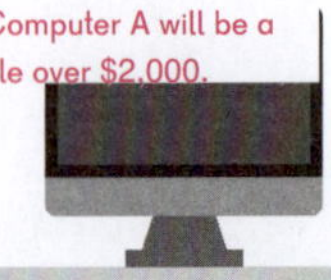

Exercise 6 ▶

Basics

1. 5,634 lights were used to decorate two stores for the holiday season.
The first store used 3,456 lights.
How many more lights did it use than the second store?

(a) Label the bar model below with the information given.
Mark the quantity that needs to be found with a question mark.

3,456

First store

Second store

5,634

?

(b) What needs to be found first?
The number of lights used by the second store.

(c) Write an expression, estimate the answer, and then solve the first step.
5,634 − 3,456 = 2,178

(d) Write an expression, estimate the answer, and solve the second step.
3,456 − 2,178 = 1,278

(e) Write a sentence with the answer to the problem.
The first store used 1,278 more lights than the second store.

Practice

Solution steps and models may vary.

2. A mall used three different colors of lights for its decorations.
It used 2,323 blue lights.
It used 736 fewer pink lights than blue lights.
It used 1,127 more purple lights than pink lights.

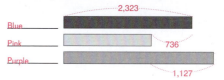

2,323

Blue

Pink   736

Purple

1,127

(a) How many purple lights did the mall use?
   2,323 − 736 = 1,587   or   1,127 − 736 = 391
   1,587 + 1,127 = 2,714   2,323 + 391 = 2,714

   2,714 purple lights

(b) How many fewer blue lights did the mall use than purple lights?
   2,714 − 2,323 = 391   or   1,127 − 736 = 391

   391 fewer blue lights

(c) How many lights did the mall use in all?
   pink lights: 2,323 − 736 = 1,587

   2,323 + 1,587 + 2,714 = 6,624

   6,624 lights

3. A 4-seat sofa set costs $1,309.
The 3-seat sofa set cost $810 less.
How much do both sets cost altogether?

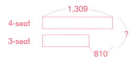

1,309

4-seat

3-seat   ?

810

   1,309 − 810 = 499
   1,309 + 499 = 1,808

Both sets cost $1,808.

4. The sum of two numbers is 9,546.
One number is 4,583.
What is the difference between the two numbers?

Students do not have to know initially which number is smaller. Steps are the same in either case.

4,583   ?

9,546

   9,546 − 4,583 = 4,963
   4,963 − 4,583 = 380

The difference between the two numbers is 380.

5. Kona and Amy have 318 coins altogether.
Ivy and Kona have 294 coins altogether.
Ivy has 124 coins.
How many coins does Amy have?

318   ?

| Kona | Amy |
| Kona | Ivy |

294   124

   294 − 124 = 170
   318 − 170 = 148

Amy has 148 coins.

6. In January, there were 4,788 migratory shore birds on an island.
In February, 3,843 birds left the island.
At the end of March, only 623 birds remained.
How many birds left the island in March?

4,788

January

February   3,843

March

623   ?

   4,788 − 3,843 = 945
   945 − 623 = 322

322 birds left in March.

Challenge

7. 30 fewer students attended East School than West School.
Then 50 students transferred from West School to East School.
How many more students attend East School than West School now?

30

East School

West School

↓

East School   50

West School   50

?

   50 − 30 = 20
   50 + 20 = 70

70 more students attend East School than West School now.

© 2017 Singapore Math Inc.   Teacher's Guide 3A Chapter 3   121

Exercise 7

Check

1 Add or subtract.

7,260 + 385	8,743 − 67	6,488 + 54
7, 2 6 0 + 3 8 5 7, 6 4 5 **R**	8, 7 4 3 − 6 7 8, 6 7 6 **T**	6, 4 8 8 + 5 4 6, 5 4 2 **H**

7,777 − 888	7,583 − 2,736	3,956 + 5,482
7, 7 7 7 − 8 8 8 6, 8 8 9 **E**	7, 5 8 3 − 2, 7 3 6 4, 8 4 7 **E**	3, 9 5 6 + 5, 4 8 2 9, 4 3 8 **T**

5,048 − 2,276	4,984 + 2,427	7,006 − 5,489
5, 0 4 8 − 2, 2 7 6 2, 7 7 2 **T**	4, 9 8 4 + 2, 4 2 7 7, 4 1 1 **L**	7, 0 0 6 − 5, 4 8 9 1, 5 1 7 **E**

Riddle: **We see it once in a year, twice in a week, but never in a day.**
What is it?
Write the letters that match the answers above to find out.

T	H	E		L	E	T	T	E	R		E
2,772	6,542	4,847	4,857	7,411	6,889	8,676	9,438	4,847	7,645	9,338	1,517

2 Use the numbers 1, 2, 3, 4, 5, and 6 to form two 3-digit numbers with the greatest sum and with the least sum.   *Addends may vary.*

	6	4	2
+	5	3	1
1,	1	7	3

	2	3	5
+	1	4	6
	3	8	1

Use the numbers 1, 2, 3, 4, 5, and 6 to form two 3-digit numbers with the greatest difference and with the least difference.

	6	5	4
−	1	2	3
	5	3	1

	4	1	2
−	3	6	5
		4	7

3 Use estimation to match equal expressions.

5,673 + 3,792	9,895 − 1,931
8,032 − 68	3,298 + 1,981
1,429 + 3,482	9,895 − 430
6,045 − 5,968	2,894 − 2,817
6,680 − 1,401	2,298 + 2,613

4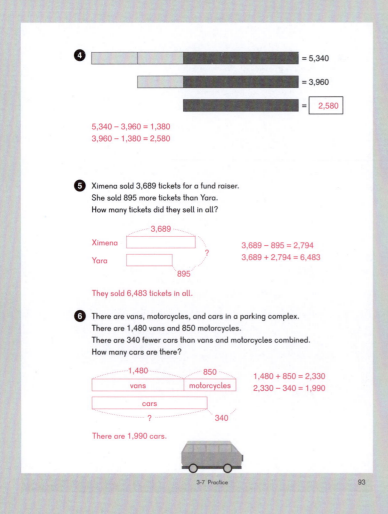

= 5,340
= 3,960
= **2,580**

5,340 − 3,960 = 1,380
3,960 − 1,380 = 2,580

5 Ximena sold 3,689 tickets for a fund raiser.
She sold 895 more tickets than Yara.
How many tickets did they sell in all?

Ximena — 3,689
Yara — ? — 895

3,689 − 895 = 2,794
3,689 + 2,794 = 6,483

They sold 6,483 tickets in all.

6 There are vans, motorcycles, and cars in a parking complex.
There are 1,480 vans and 850 motorcycles.
There are 340 fewer cars than vans and motorcycles combined.
How many cars are there?

1,480 — vans   850 — motorcycles
cars — ? — 340

1,480 + 850 = 2,330
2,330 − 340 = 1,990

There are 1,990 cars.

Challenge

7 Each shape represents a digit.
What are the digits?

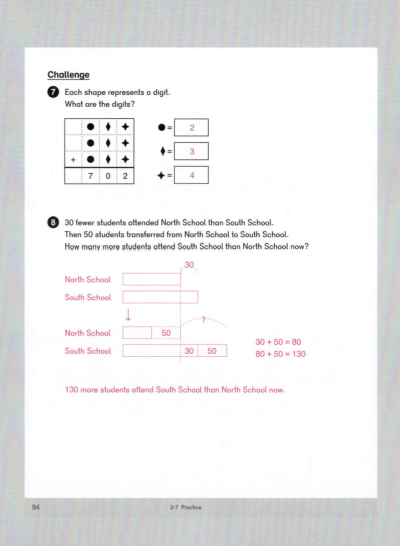

●	◆	✦
●	◆	✦
+ ●	◆	✦
7	0	2

● = **2**
◆ = **3**
✦ = **4**

8 30 fewer students attended North School than South School.
Then 50 students transferred from North School to South School.
How many more students attend South School than North School now?

North School
South School — 30

North School — 50
South School — 30 | 50 — ?

30 + 50 = 80
80 + 50 = 130

130 more students attend South School than North School now.

Suggested number of class periods: 11–12

	Lesson	Page	Resources		Objectives
	Chapter Opener	p. 129	TB:	p. 97	Review multiplication concepts.
1	Looking Back at Multiplication	p. 130	TB: WB:	p. 98 p. 95	Understand the meaning of multiplication. Understand that reversing the order of the factors results in the same product.
2	Strategies for Finding the Product	p. 133	TB: WB:	p. 102 p. 98	Use known facts to derive unknown facts. Apply the properties of operations as strategies to multiply.
3	Looking Back at Division	p. 136	TB: WB:	p. 105 p. 101	Understand sharing and grouping situations as division. Relate division to multiplication. Use multiplication facts to solve division problems.
4	Multiplying and Dividing with 0 and 1	p. 139	TB: WB:	p. 109 p. 104	Multiply a number by 0. Multiply and divide a number by 1.
5	Division with Remainders	p. 141	TB: WB:	p. 113 p. 107	Understand division with a remainder. Divide one-digit and two-digit numbers with a remainder.
6	Odd and Even Numbers	p. 144	TB: WB:	p. 116 p. 111	Understand odd and even numbers.
7	Word Problems — Part 1	p. 146	TB: WB:	p. 120 p. 114	Represent multiplication and division situations with bar models and determine the correct operation.
8	Word Problems — Part 2	p. 149	TB: WB:	p. 124 p. 116	Solve one-step and two-step word problems involving comparison situations with multiplication and division.
9	2-Step Word Problems	p. 152	TB: WB:	p. 128 p. 119	Solve two-step word problems involving all four operations.
10	Practice	p. 155	TB: WB:	p. 132 p. 123	Practice topics from the chapter.
	Review 1	p. 157	TB: WB:	p. 135 p. 127	Review content from Chapters 1 through 4.
	Workbook Solutions	p. 159			

In **Dimensions Math® 2A** and **2B**, students learned to:

- Understand multiplication and division as working with equal groups.
- Understand and master multiplication and division tables of 2, 3, 4, 5, and 10.
- Interpret multiplication and division models for 1-step word problems.

In this chapter, students formalize their knowledge of multiplication and division. New concepts include:

- Multiplying and dividing with 0 and 1
- Remainders
- Odd and even numbers
- Representing and interpreting multi-step problems for the four operations

Multiplication and Division

Multiplication and division can be understood using the part-whole concept, however, as the relationship is multiplicative, number bonds are not used. A number bond for $4 \times 3 = 12$ is possible, but when students move to problems such as 10×3 or 10 threes, number bonds are not helpful and bar models are used instead.

Since 3×10 and 10×3 result in the same product, students can write the factors in any order. Multiplication is commutative and it may be easier for students to skip count by tens than by threes.

Lessons 1 through 3 review multiplication facts and related division facts, arrays, and the terms **times** and **product**. Students will practice strategies for finding facts they don't know and create flash cards for

multiplication facts. The term **quotient** is introduced in Lesson 3.

Multiplying and dividing with 0 and 1 is covered in Lesson 4. Many students discover these properties independently:

- The product of any number and 0 is always 0.
- The product of any number and 1 is that number.
- 0 groups of anything is 0.
- The quotient when a number is divided by 0 is undefined. There is no way to turn any number of things into 0 groups.
- A number divided by 1 is itself.
- A number divided by itself is 1.

Students should continue to practice multiplication and division facts until they know them by memory.

Students will learn the 10 remaining facts in **Dimensions Math® 3B Chapter 8: Multiplying and Dividing with 6, 7, 8, and 9.**

Remainders

Division with a remainder is a procedure for handling an arithmetic problem where the dividend is not a multiple of the divisor.

For example, $16 \div 5$:
16 apples cannot be put into equal groups of 5. There will be 3 groups of 5, with 1 left over.

Division with remainder is shorthand for, "How many equal groups, and how many left over?"

At this level, students should learn that

$16 = 5 \times 3 + 1$

where 5 is the quotient and 1 is the remainder.

Note: Often, division with remainder problems are written in the notation:

$16 \div 3 = 5 \text{ R } 1$

Using the = sign in an equation with a remainder is technically incorrect. Compare 16 ÷ 3 = 5 R 1 to the following:

21 ÷ 4 = 5 R 1

The answers appear to be the same. However, 16 ÷ 3 does not equal 21 ÷ 4 and when the actual answer is given as a fraction, we see they are not the same.

$$16 \div 3 = 5\frac{1}{3} \quad \text{and} \quad 21 \div 4 = 5\frac{1}{4}$$

Students will not be taught to understand that remainders can be expressed as fractions until **Dimensions Math® 4**. The term remainder or, in this example, "5 R 1," will occasionally be used, but this notation will not be used with a math equation.

The abbreviated R notation is used in the answer overlay for space considerations.

Even Numbers

Even numbers are introduced as numbers that can be divided by 2 with no remainder. A common student misconception is that 0 is both odd and even. Odd numbers have a remainder of 1 when divided by 2. As such, 0 is even.

Investigations of products of even and odd numbers are in Lesson 5.

Bar Models

Lessons in this chapter further develop the use of bar models to represent and interpret multiplication and division word problems. Lesson 9 introduces multi-step problems for the four operations.

Students benefit from model drawing because it provides a strategy to visualize the problem pictorially, and determine an approach to solving it.

The computations in these bar models use facts for 2, 3, 4, 5, and 10 that students learned in

Dimensions Math® 2A and **2B**. The focus is on the problem solving methods.

In this chapter, students will be asked to draw the models. As they solve harder problems they will begin to appreciate the usefulness of the model drawing strategy. Students will be expected to:

- Draw a bar model for each problem.
- Write expressions and solve them for each computation step needed to solve the problem.
- Put the answer in a complete sentence, for example, "Alex had $52 at first." This is to encourage students to not merely solve the problem correctly, but to solve the correct problem.

Bar models are used for word problems throughout the **Dimensions Math®** series. Working through word problems is how students realize the importance and applications of mathematics.

Part-Whole Models extend the understanding of multiplicative relationships. Each part is an equal group and is considered a "unit."

There are 3 children. Each child has 8 pencils. How many pencils do they have in all?

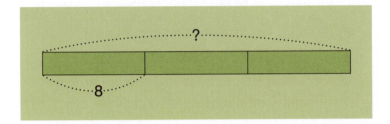

1 unit ⟶ 8 pencils
3 units ⟶ 3 × 8 = 24

There are 24 pencils in all.

In **sharing division**, the whole is shared evenly among the parts.

Mrs. Turner has 24 pencils. She gives each of her 3 students an equal number of pencils. How many pencils does each student receive?

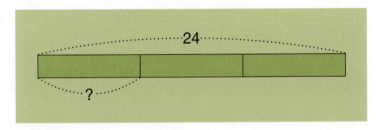

3 units ⟶ 24 pencils
1 unit ⟶ 24 ÷ 3 = 8

Each student receives 8 pencils.

The model for **grouping division** can be harder for students to grasp:

Mrs. Turner has 24 pencils. She gives 3 pencils to each of her students. How many students receive pencils?

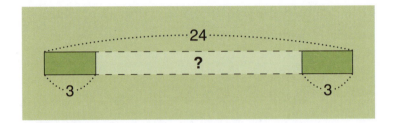

1 unit ⟶ 3 pencils
? units ⟶ ? × 3 = 24

24 ÷ 3 = 8

8 students received pencils.

The reason for the dotted lines in the model and an equal group on each end is that we are visualizing the bar is filled with groups, each containing 3 pencils, but we do not know how many groups of pencils there are.

Comparison Models are useful to compare quantities. They are particularly useful in representing multi-step problems.

There are 20 red pencils. There are 4 times as many red pencils as yellow pencils. How many yellow pencils are there?

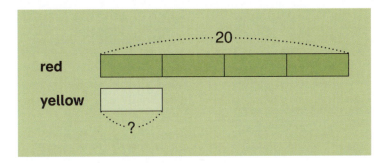

4 units ⟶ 20 red pencils
1 unit ⟶ 20 ÷ 4 = 5

There are 5 yellow pencils.

From this same model, students can also find:

(a) How many more red pencils than yellow pencils were there?

(b) How many pencils in all?

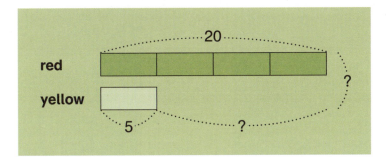

(a) Students could subtract: 20 − 5 = 15

Or multiply: 1 unit ⟶ 5
            3 units ⟶ 3 × 5 = 15

There are 15 more red pencils than yellow pencils.

(b) Students could add: 20 + 5 = 25
Or multiply: 5 units ⟶ 5 × 5 = 25

There are 25 pencils in all.

Materials

- Counters
- Dice
- Die with modified sides: 0 in place of 6
- Dry erase markers
- Dry erase sleeves
- Half-inch graph paper
- Multiplication and division fact cards for 0 to 5
- Paper plates with the center cut out
- Plastic bags
- Recording sheet
- Strips of paper
- Two-color counters
- Whistle
- Whiteboards

Blackline Masters

- Centimeter Graph Paper
- Kaboom Cards
- Missing Products 4–2
- Number Cards
- Secret Math Message Decoder

Storybooks

- *The Multiplying Menace Divides* by Paul Calvert
- *The Great Divide* by Dayle Ann Dodds
- *The Doorbell Rang* by Pat Hutchins
- *A Remainder of One* by Elinor J. Pinczes
- *One Hundred Angry Ants* by Elinor J. Pinczes

Activities

Games and activities included in this chapter provide multiplication and division practice, and can be adapted for other numbers in the multiplication and division tables.

Activities can be used after students complete the **Do** questions, or any time additional practice is needed. Students should know their multiplication facts for 0 through 5, and for 10, from memory by the end of the chapter.

Chapter Opener

Objective

- Review multiplication concepts.

Using the **Chapter Opener** illustration, have students create and share multiplication and division word problems about the crafts.

Possible student answers:

- The caterpillar uses 7 pompoms. How many pompoms do I need to make 4 caterpillars? 7 × 4
- Each spider needs 2 googly eyes. How many spiders can I make with 18 googly eyes? 18 ÷ 2
- I need 4 pipe cleaners to make each spider. How many pipe cleaners do I need to make 6 spiders? 4 × 6

Teachers should assess students' prior knowledge of multiplication facts for 2, 3, 4, 5, and 10.

This lesson may continue straight to **Lesson 1: Looking Back at Multiplication**.

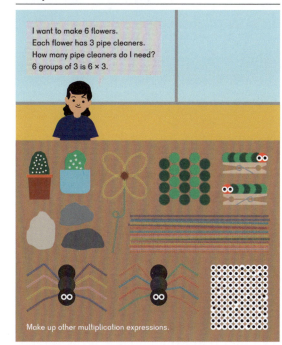

Chapter 4

Multiplication and Division

I want to make 6 flowers.
Each flower has 3 pipe cleaners.
How many pipe cleaners do I need?
6 groups of 3 is 6 × 3.

Make up other multiplication expressions.

97

Lesson 1 Looking Back at Multiplication

Objectives

- Understand the meaning of multiplication.
- Understand that reversing the order of the factors results in the same product.

Lesson Materials

- Counters

Think

Pose the **Think** problem and give students counters. Allow students adequate time to find multiplication expressions.

Discuss student strategies.

Learn

Multiplication with whole numbers can be solved using repeated addition.

Dion and Alex discuss the different ways of considering how many equal groups there are.

Students should be familiar with the × sign and term **product**. The term **multiplied by** is introduced as an additional way of thinking about multiplication.

Have students compare their methods from **Think** with the methods shown in the textbook.

Students should note that it doesn't matter which number represents the number of groups and which represents the number in each group, the answer is the same. Both 4 groups of 6 and 6 groups of 4 yield the same answer.

Since:

$6 + 6 + 6 + 6 = 4 + 4 + 4 + 4 + 4 + 4$

Then:

$6 × 4 = 4 × 6$.

Students should understand why these sums are the same.

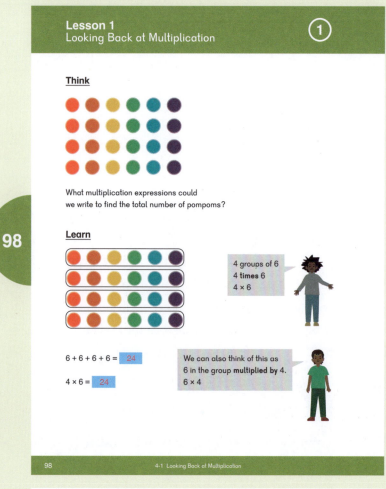

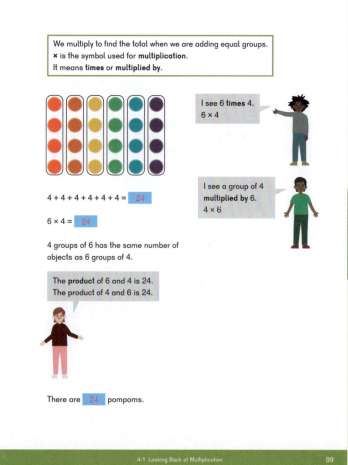

98

99

    © 2017 Singapore Math Inc.

Do

①–② These problems with the ants and leaves and the pipe cleaners illustrate the fact that although the (a) and (b) pictures differ, the product of 3 and 6 (**①**), and 9 and 4 (**②**) remains the same.

③ The garlic head array can be seen as either 3 rows of 10 or 10 columns of 3.

Take this opportunity to review multiplication facts for 2 to 5.

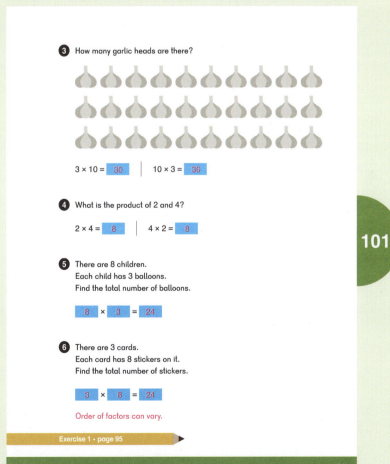

Activities

▲ Multiplication Wheels

Materials: Paper plates with the center cut out, whiteboards, dry erase markers

Create several multiplication wheels with the numbers 1 to 10 in random order as "spokes" along the edge of the paper plates.

Have students lay the wheels on their whiteboards and write the number they are multiplying by in the center of the wheel. (In the example below, it is 5.)

Students multiply the number on the spoke and the number in the center, and write the product on the whiteboard, outside of the wheel.

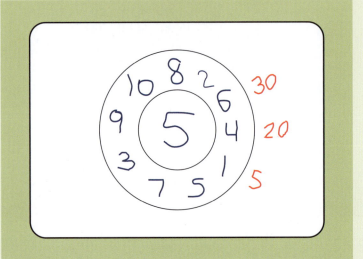

▲ Fences

Materials: Centimeter Graph Paper (BLM) or half-inch graph paper in dry erase sleeves, 2 dice, 2 colors of dry erase markers

Students take turns rolling the dice. They fence in land on the game board by outlining an array with the same number of spaces as the dice roll.

The players write two multiplication equations on their newly acquired land.

In the example shown below:

- Player One (blue) rolled 5 and 4 and fenced in a 5 × 4 array.
- Player Two (red) rolled 2 and 6 and fenced in a 2 × 6 array.

Play stops when a player can't fit his array on the board. Each player adds up the total amount of land (or boxes) he has fenced in. The player with the most land is the winner.

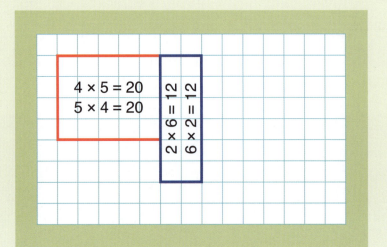

◄ **Exercise 1 • page 95**

Lesson 2 Strategies for Finding the Product

Objectives

- Use known facts to derive unknown facts.
- Apply the properties of operations as strategies to multiply.

Lesson Materials

- Two-color counters
- Missing Products 4—2 (BLM)

Think

Pose the **Think** problem. Provide students with two-color counters and Missing Numbers 4—2 (BLM).

Ask students what strategies they could use to find the facts that they do not know. Possible student answers:

- I could count the counters.
- I could just count on more from the last answer. (Alex's strategy)
- I notice patterns in the columns that help find the answers.
- I can combine facts that I know. (Sofia and Mei's strategy)

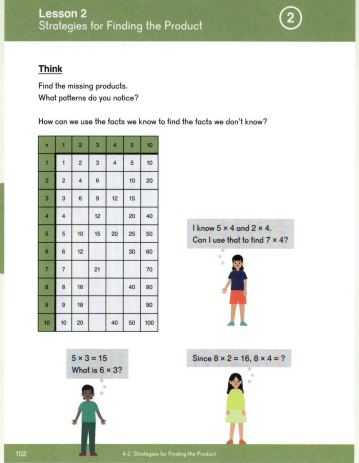

Lesson 2
Strategies for Finding the Product ②

Think

Find the missing products.
What patterns do you notice?

How can we use the facts we know to find the facts we don't know?

×	1	2	3	4	5	10
1	1	2	3	4	5	10
2	2	4	6		10	20
3	3	6	9	12	15	
4	4		12		20	40
5	5	10	15	20	25	50
6	6	12			30	60
7	7		21			70
8	8	16			40	80
9	9	18				90
10	10	20		40	50	100

I know 5 × 4 and 2 × 4.
Can I use that to find 7 × 4?

5 × 3 = 15
What is 6 × 3?

Since 8 × 2 = 16, 8 × 4 = ?

102   4-2 Strategies for Finding the Product

Learn

Discuss the solutions the 3 friends used to find their product.

In each case, the friends are deriving new facts from known facts.

Students may see other ways of finding facts for a number using other facts that are not included in the text:

- For 6 × 3, students could think 6 is double threes, so 6 × 3 is two groups of 3 × 3, or 9 + 9 as shown in **Do ❷**.
- 7 × 4 is the same as (4 × 4) + (3 × 4), or 16 + 12.
- For 8 × 4, students could think of 8 as 4 + 4 , so 8 × 4 is two groups of 4 × 4, or 16 + 16.

Dion reminds students that if they know the fact one way, changing the order of the numbers results in the same product.

Note: Students are using a visual representation to learn how to apply operations for multiplication without learning them formally.

Students will then be able to use these operations to help them derive the remaining multiplication facts they have not yet learned, or to help them quickly compute any they have forgotten.

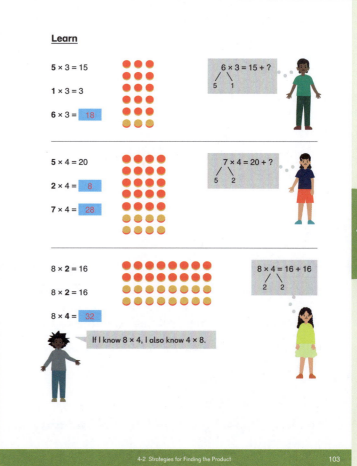

Do

1 — **4** Have students discuss the strategies shown for each problem.

1 Emma is reviewing the idea that 9 groups of a number is 1 group less than 10 groups of that number.

Activity

▲ **Greatest Product**

Materials: 4 sets of Number Cards (BLM) 0 to 10, recording sheet, die with modified sides: 0 in place of 6

Shuffle the Number Cards (BLM) and place them facedown in the middle of the players.

On each turn, a player draws a Number Card (BLM) and rolls the die. That player multiplies the number on the Number Card (BLM) and the die and records the product.

When each player has had a turn, players compare their products. The player with the greatest product gets a point. The first player to get 10 points is the winner.

Exercise 2 • page 98

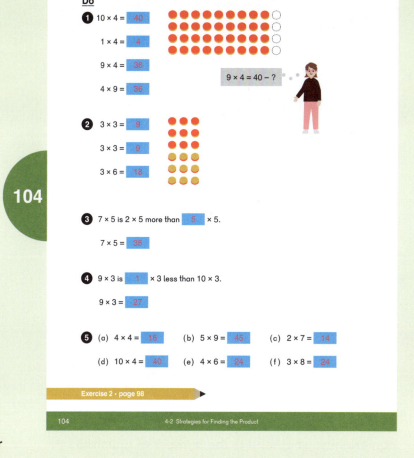

Lesson 3 Looking Back at Division

Objectives

- Understand sharing and grouping situations as division.
- Relate division to multiplication.
- Use multiplication facts to solve division problems.

Lesson Materials

- Plastic bags
- Counters

Think

Provide students with plastic bags and 15 counters each and pose the **Think** problem.

Have students use the counters to show the different types of division as either sharing or grouping.

Discuss student strategies for solving the problem.

Students should notice that the answer is the same numerically, whether it is the number of counters (a) or the number of bags (b).

Learn

Have students discuss the two different division situations in **Think**, and how they are written as equations.

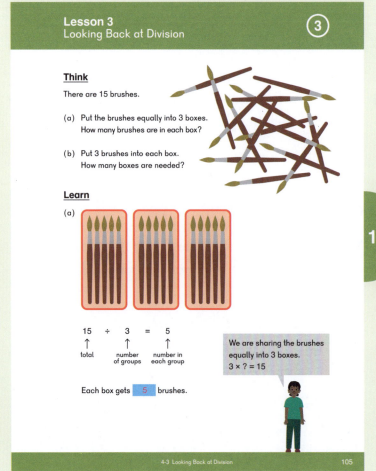

Although there are two different problem situations, both situations can be represented with division expressions. The quotient is the same for both problem situations, but the meaning of each quotient is different.

Discuss the terms with students. They learned the terms **divided by**, **sharing**, and **grouping** in **Dimensions Math® 2A**, so these should be review.

Alex introduces the term **quotient** as the answer to a division equation.

Provide students with additional examples of sharing and grouping division and have them show the situations with counters. Distribute more counters so students can come up with more examples.

Ask students to come up with their own examples of sharing and grouping situations:

* We had 12 students and put them into 2 teams for soccer.
* We invited 16 guests and had to set up card tables. Each table held 4 people.
* We made up bags of towels to give to the dog shelter. Each bag had to have 3 towels.

Do

Have struggling students use counters to work through the problems. Most students should be working with the images in the textbook.

❷ – ❸ These problems illustrate the four related multiplication and division equations that can be found.

❹ Each problem relates a multiplication fact to a division equation.

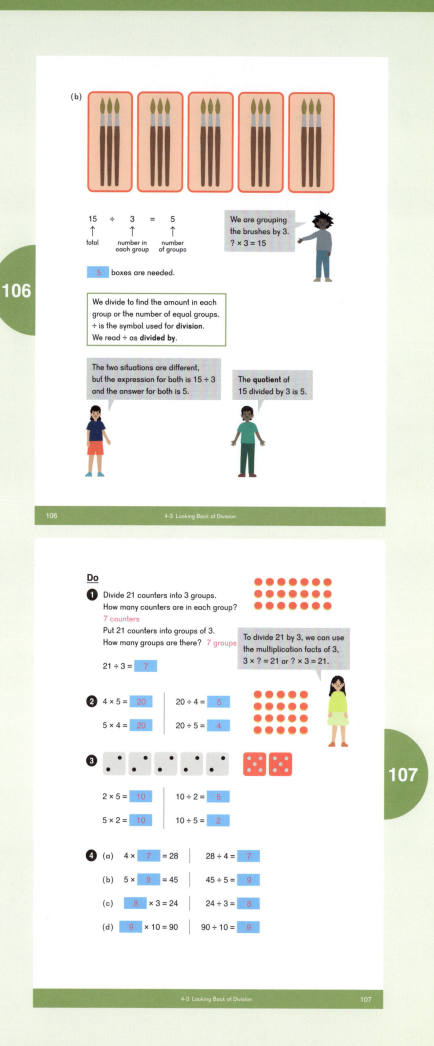

5 Students may think of related multiplication facts to help them find the quotients.

6 Using related multiplication facts is a good strategy for this problem. If Ximena has one nickel, she has 5 cents...

- 1 × 5 = 5
- 2 × 5 = 10
- 3 × 5 = 15
- ? × 5 = 45

Students may note that this is a grouping division problem: 45 cents put into groups of 5 cents. How many groups of 5 cents can we make?

Nine groups of 5 cents is the same value as 9 nickels.

Activity

▲ Student Sorts

Materials: Whistle

Have students put themselves into groups.

- Line up in 2 rows.
- Put yourselves into 5 groups. How many are in each group? How many in all?
- Put yourselves into groups of 5. How many groups are there? Are there any students left over?
- Get into groups of 5 × 5 − 5.

Try blowing a whistle a certain number of times and have students get into groups of that number and link arms.

Students without a group get to help the teacher decide the next number given to the other students.

Exercise 3 • page 101

108

Lesson 4 Multiplying and Dividing with 0 and 1

Objectives

- Multiply a number by 0.
- Multiply and divide a number by 1.

Lesson Materials

- Counters

Think

Pose the **Think** problem. Provide students with counters to show the division examples in **Think**. Discuss students' ideas.

Learn

The purpose of this lesson is to extend understanding of the properties of division informally by helping students see what happens when they multiply and divide a number by 0 or 1.

Discuss Alex and Emma's statements. Students know that division facts have related multiplication facts. Point out to them that we cannot find 4 ÷ 0 because there is no number multiplied by 0 that gives a product of 4.

$$4 \div 0 = \boxed{} \longrightarrow \boxed{} \times 0 = 4$$

Do

❶ The pattern in (a−c) shows that as the number of birds on each branch decreases by one, the product goes down by 3.

(d−e) The number of branches decreases by 1, but since there are no birds, the answer is always 0 birds on the branch. In this problem students can see that 0 groups of 0 also results in a product of 0, leading to Mei's comment.

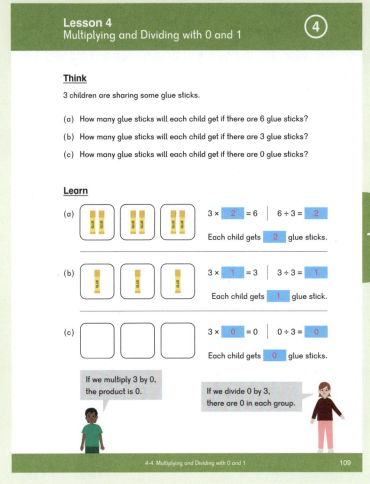

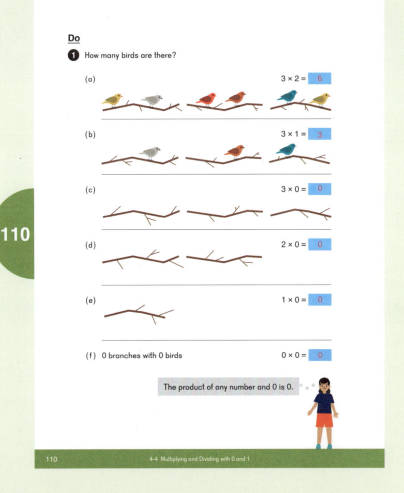

2 Students may also note that 6 counters, 1 in each group, is 6 counters grouped by 1. Have them write the equation for 6 counters in 6 groups. This will be further developed in **3**.

Dion points out that we can't put objects into 0 groups.

3 Sofia guides students to a property of division: Any number divided by itself is 1.

Students do not need to memorize these properties of multiplication and division. They should divide counters to understand what it means to divide a number by 1 and by itself.

Exercise 4 · page 104

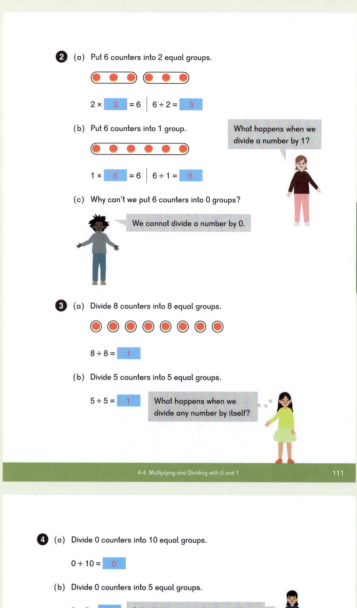

2 (a) Put 6 counters into 2 equal groups.

2 × [3] = 6 | 6 ÷ 2 = [3]

(b) Put 6 counters into 1 group.

What happens when we divide a number by 1?

1 × [6] = 6 | 6 ÷ 1 = [6]

(c) Why can't we put 6 counters into 0 groups?

We cannot divide a number by 0.

3 (a) Divide 8 counters into 8 equal groups.

8 ÷ 8 = [1]

(b) Divide 5 counters into 5 equal groups.

5 ÷ 5 = [1] *What happens when we divide any number by itself?*

111

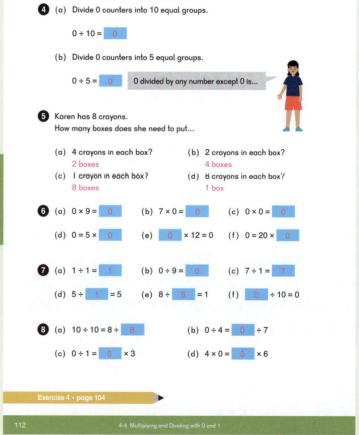

4 (a) Divide 0 counters into 10 equal groups.

0 ÷ 10 = [0]

(b) Divide 0 counters into 5 equal groups.

0 ÷ 5 = [0] *0 divided by any number except 0 is...*

5 Karen has 8 crayons.
How many boxes does she need to put...

(a) 4 crayons in each box? (b) 2 crayons in each box?
2 boxes   4 boxes

(c) 1 crayon in each box?   (d) 8 crayons in each box?
8 boxes   1 box

6 (a) 0 × 9 = [0]   (b) 7 × 0 = [0]   (c) 0 × 0 = [0]

(d) 0 = 5 × [0]   (e) [0] × 12 = 0   (f) 0 = 20 × [0]

7 (a) 1 ÷ 1 = [1]   (b) 0 ÷ 9 = [0]   (c) 7 ÷ 1 = [7]

(d) 5 ÷ [1] = 5   (e) 8 ÷ [8] = 1   (f) [0] ÷ 10 = 0

8 (a) 10 ÷ 10 = 8 ÷ [8]   (b) 0 ÷ 4 = [0] ÷ 7

(c) 0 ÷ 1 = [0] × 3   (d) 4 × 0 = [0] × 6

Exercise 4 · page 104

112

Objectives

- Understand division with a remainder.
- Divide one-digit and two-digit numbers with a remainder.

Lesson Materials

- Counters

Think

Provide students with 14 counters and have them work through the **Think** problem.

Remind students that when we divide, we need to make equal groups.

Learn

Discuss the **Think** problem with students and have them share their solutions. They should note that there are some counters left over that cannot be put into equal groups of 3.

Mei explains that we refer to the leftover counters as a **remainder**. The remainder has not been divided by 3.

Sofia is estimating by using a known multiplication fact.

Discuss the use of both the **R** notation as well as the meaning of the expression: $4 \times 3 + 2$, or 4 groups of 3 pipe cleaners with 2 left over.

The remainder will always be less than the divisor because if the remainder is equal to or greater than the divisor, another group can be made. This is a common error students make (for example, $14 \div 3 =$ 3 R 5) and will be further discussed in ❸.

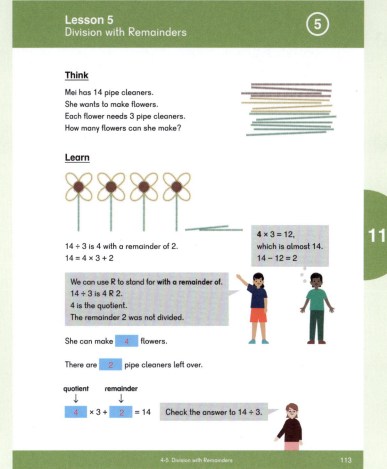

Lesson 5
Division with Remainders ⑤

Think

Mei has 14 pipe cleaners.
She wants to make flowers.
Each flower needs 3 pipe cleaners.
How many flowers can she make?

Learn

$14 \div 3$ is 4 with a remainder of 2.
$14 = 4 \times 3 + 2$

$4 \times 3 = 12$, which is almost 14.
$14 - 12 = 2$

We can use R to stand for with a remainder of.
$14 \div 3$ is 4 R 2.
4 is the quotient.
The remainder 2 was not divided.

She can make ⬚4⬚ flowers.

There are ⬚2⬚ pipe cleaners left over.

quotient   remainder
↓   ↓
⬚4⬚ × 3 + ⬚2⬚ = 14   Check the answer to $14 \div 3$.

4-5 Division with Remainders   113

Do

1 Dion thinks the remainder has to be less than the number we are dividing by because if it is equal to or greater then there is a another group that can be formed. This will be revisited in ❸.

2 Emma estimates using a known multiplication fact. She splits 24 into 20, a number that she knows is a fact for 5.

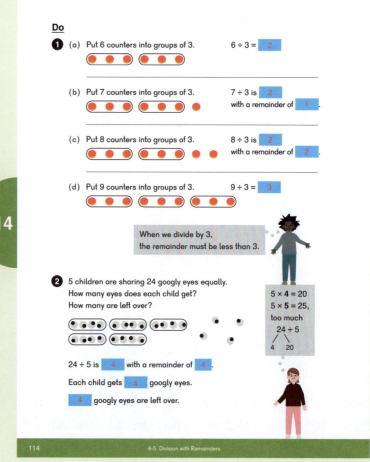

114

③ Repeat this activity with counters for students who struggle:

- 12 ÷ 4 is 3 with no remainder.
- 13 ÷ 4 is 3 with a remainder of 1.
- 14 ÷ 4 is 3 with a remainder of 2.
- 15 ÷ 4 is 3 with a remainder of 3.
- 16 ÷ 4 is 4 with no remainder.
- 17 ÷ 4 is 4 with a remainder of 1.
- 18 ÷ 4 is 4 with a remainder of 2.
- 19 ÷ 4 is 4 with a remainder of 3.
- 20 ÷ 4 is 5 with no remainder.

Discuss Alex's thought about remainders. Students should note that they should not have a remainder equal to or greater than the number they are dividing by.

⑤ **Note:** Although 46 ÷ 10 is 4 remainder 6, they will need 5 vans to hold all of the people.

Activity

▲ Remainders

Materials: 45 counters, die with modified sides: 0 in place of 6

Player One rolls the die and divides her counters by the number on the die.

For example, Player One rolls a 2. She divides the counters into 2 equal groups with 1 left over. She keeps the remaining counter and play continues with 44 counters.

Player Two rolls a 5 and divides the remaining 44 counters by 5. She has 8 groups of 5, and keeps the remaining 4 counters. Play continues with 40 counters.

When no more divisions can be made, the game is over. The player with the most counters is the winner.

Exercise 5 • page 107

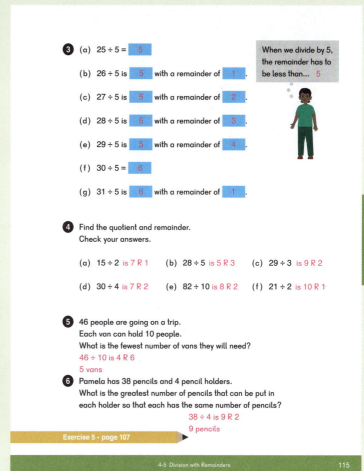

③ (a) 25 ÷ 5 = **5**

(b) 26 ÷ 5 is **5** with a remainder of **1**.

(c) 27 ÷ 5 is **5** with a remainder of **2**.

(d) 28 ÷ 5 is **5** with a remainder of **3**.

(e) 29 ÷ 5 is **5** with a remainder of **4**.

(f) 30 ÷ 5 = **6**

(g) 31 ÷ 5 is **6** with a remainder of **1**.

When we divide by 5, the remainder has to be less than... 5

④ Find the quotient and remainder. Check your answers.

(a) 15 ÷ 2 is 7 R 1 (b) 28 ÷ 5 is 5 R 3 (c) 29 ÷ 3 is 9 R 2

(d) 30 ÷ 4 is 7 R 2 (e) 82 ÷ 10 is 8 R 2 (f) 21 ÷ 2 is 10 R 1

⑤ 46 people are going on a trip.
Each van can hold 10 people.
What is the fewest number of vans they will need?
46 ÷ 10 is 4 R 6
5 vans

⑥ Pamela has 38 pencils and 4 pencil holders.
What is the greatest number of pencils that can be put in each holder so that each has the same number of pencils?
38 ÷ 4 is 9 R 2
9 pencils

Exercise 5 • page 107

4-5 Division with Remainders   115

Lesson 6 Odd and Even Numbers

Objective

- Understand odd and even numbers.

Lesson Materials

- Counters

Think

Provide students with counters and allow them adequate time to work on the **Think** problem. Have them write equations for each problem.

Discuss student strategies for solving the problem.

Learn

Discuss Sofia's comments. A number that cannot be divided evenly by 2 is called an odd number.

The number line with the blue and green dots provides a nice visual for students to see representations of even and odd numbers. Students should see that for the odd numbers, there is always one dot that does not have a partner.

Discuss Dion's comment that 0 is an even number.

An even number is a multiple of 2. This explains why 0 is even, specifically, 0 × 2 gives a product that is a multiple of 2.

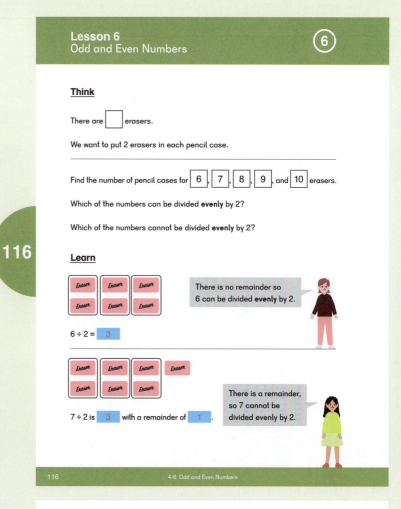

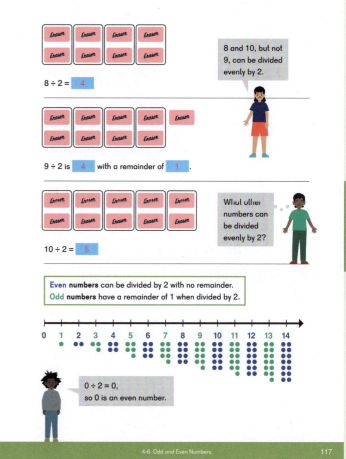

Do

5 Discuss Mei's thought on the products of odd and even numbers. When students multiply an odd number by an even number, the result is even. Only multiplying an odd number by an odd number will result in a product that is an odd number.

★ Have students decide if there are additional rules for adding and subtracting odd and even numbers. For example, will an even − odd = even or odd?

Activity

● **Odds & Evens**

Students pair up and one of them chooses to be "odds" while the other is "evens." Similar to Rock, Paper, Scissors, both players say, "One, two, three, shoot!" On "shoot," players shoot out either 1 or 2 fingers.

The sum total of fingers is either odd or even. If the result is odd, then the player who chose odds is the winner. If the sum is even, the player who chose evens is the winner.

Exercise 6 • page 111

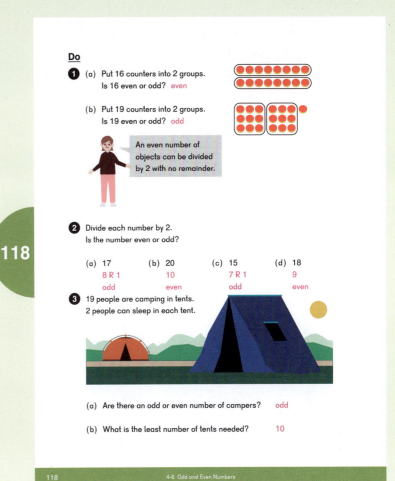

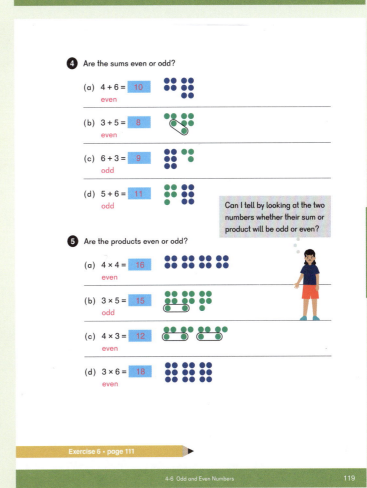

Objective

- Represent multiplication and division situations with bar models and determine the correct operation.

Think

Pose the problems in **Think** and have students draw a model for each. Have students share how they solved the problems.

Students may draw models that do not look like the ones in **Learn**.

For example...

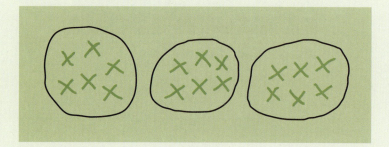

is an alternative way of showing 3 pots with 6 rocks in each pot.

Discuss Alex's questions about how the problems are similar and how they are different.

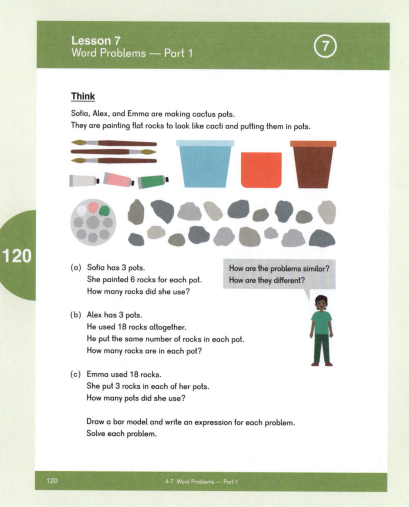

Lesson 7
Word Problems — Part 1 ⑦

Think

Sofia, Alex, and Emma are making cactus pots.
They are painting flat rocks to look like cacti and putting them in pots.

(a) Sofia has 3 pots.
She painted 6 rocks for each pot.
How many rocks did she use?

How are the problems similar?
How are they different?

(b) Alex has 3 pots.
He used 18 rocks altogether.
He put the same number of rocks in each pot.
How many rocks are in each pot?

(c) Emma used 18 rocks.
She put 3 rocks in each of her pots.
How many pots did she use?

Draw a bar model and write an expression for each problem.
Solve each problem.

120   4-7 Word Problems — Part 1

Learn

Have students discuss the bar models in **Learn** and compare their own models with the ones in the textbook.

Multiplication and division have parts that are equal in amount. We call these units. Discuss Dion's comment about units as equal groups. Have students notice the way the steps are written in the textbook. The arrow means "is" or "represents." For example, "1 unit represents 6 rocks, so 3 units represents 3 × 6, or 18, rocks."

In (b), Sofia reminds students that they know the total and can divide to find the unit.

Students should see that all units represent the same amount and only one unit needs to be labeled.

(c) can be challenging for students to understand. Discuss:

- If 1 pot has 3 rocks, how many pots are needed to use all 18 rocks?
- The reason we have the dotted lines and an equal group on each end is that we are visualizing the bar filled with groups, each containing 3 rocks, but we do not know how many groups there are yet.

Ask students why the question marks on each bar model are in their given locations.

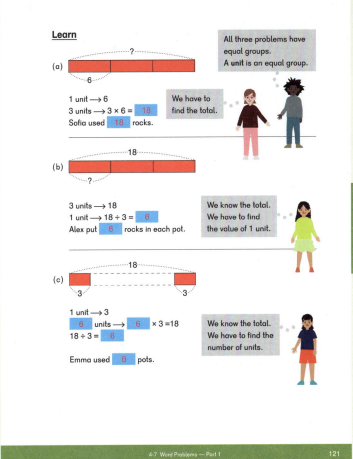

Do

1 Have students write expressions for the bar models and solve.

3 Help students see that we are determining how many groups are needed, not how many are in a group.

5 Ask students:

- What do we need to find?
- Do we know the value of one of the units?
- Do we know the whole or total?

The task is to:

- Draw a bar model for each problem.
- Write an expression and solve for each problem.
- Solve each problem. An answer will be in a complete sentence, for example:
  (a) There are 40 decorations in all.

Students may use units and arrows to show their thinking in arriving at an expression:

(a)   1 unit ⟶ 10 decorations
      4 units ⟶ 4 × 10 = 40 decorations

   "If 1 unit is 10 decorations, then 4 units is 4 times as many decorations."

(b)   1 unit ⟶ 2 feet of yarn
      ? units ⟶ ? × 2 = 20 feet of yarn

   "If 1 unit is 2 feet, how many units are needed to make 20 feet?"

(c)   3 units ⟶ 27 pieces of ribbon
      1 unit ⟶ 27 ÷ 3 = 9 pieces of ribbon

   "3 units is 27 pieces of ribbon, then 1 unit is 27 divided by 3 pieces of ribbon."

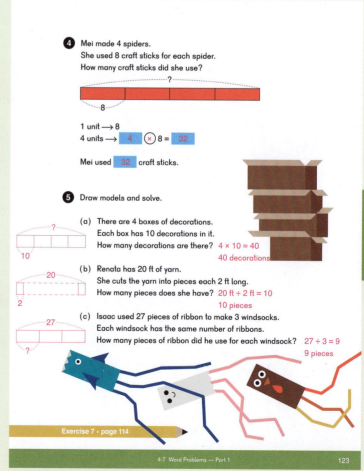

Exercise 7 • page 114

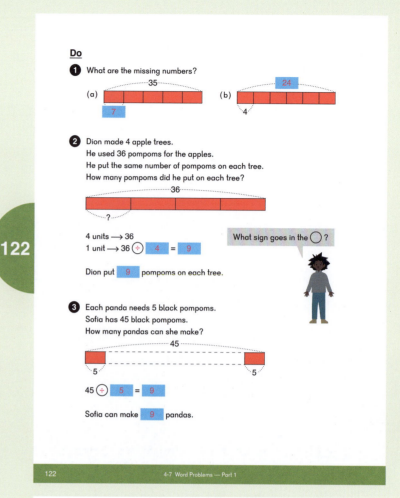

Objective

- Solve one-step and two-step word problems involving comparison situations with multiplication and division.

Think

Pose the problems in **Think** and have students discuss how they can draw a bar model for the problems.

Ask students:

- How are these problems similar to the ones we did yesterday? (They are about rocks. They are about multiplying and dividing.)
- How are they different? (We are comparing.)

124

Lesson 8
Word Problems — Part 2   ⑧

Think

Mei, Dion, and Sofia are hunting for two sizes of rocks, large and small, for an art project.

(a) Mei found 32 small rocks.
She found 4 times as many small rocks as large rocks.
How many large rocks did she find?

(b) Dion found 8 large rocks.
He found 4 times as many small rocks as large rocks.
How many rocks did he find in all?

(c) Sofia found 24 more small rocks than large rocks.
She found 4 times as many small rocks as large rocks.
How many large rocks did she find?

Draw a bar model and write an expression for each problem.
Solve each problem.

124   4-8 Word Problems — Part 2

Learn

Have students discuss the bar models in **Learn**. Note that the bars are drawn the same for all three problems, however the quantities given and questions differ.

For (a), ask students to think about what 4 times as many might look like. Note that we are comparing the number of large rocks to small rocks and will need to represent both quantities.

Repeat for Dion and Sofia's problems.

Ask students how the models represent the quantities in each problem:

- Does the friend have more small rocks or large rocks? How is that shown on the model?
- What is 1 unit? The small rocks or the large rocks?
- How can we find how many more large rocks (or small rocks) there are?

For (b) only:

- Why do we divide by 3 when there are 4 units?
- How is this model different than the previous two problems?

Do

❶ Have students write the expressions and solve.

❷ Ask students:

- Why is the model drawn and labeled the way that it is in the textbook?
- Could the bar that represents monkeys be on top and the bar for snakes be on the bottom?

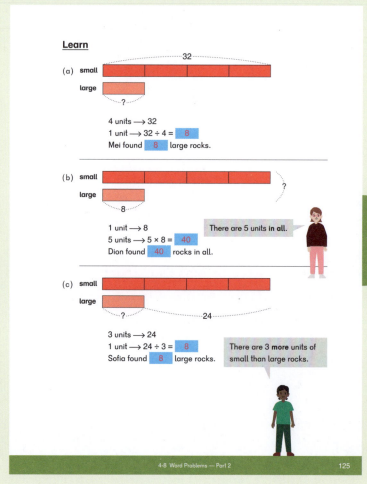

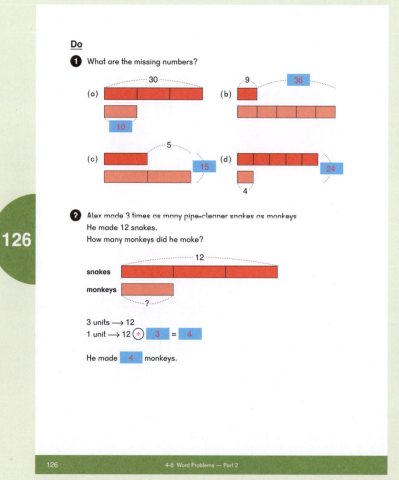

125

126

3 Ask students:

- Does it matter if the bar representing paint is on the top or bottom?
- Why does the solution show 4 units when there are 5 units of paint?

4 Ask students:

- What do we need to find?
- Are we comparing?
- Do we know the whole or total?
- What is the value of 1 unit?
- How many units are there in all?

Remind students that the task is to:

- Draw a bar model for each problem.
- Write an expression for each problem.
- Solve each problem.
- State the solution in a complete sentence. For example, for (a), "Susma has 3 sticks of red clay."

Students may use units and arrows to show their thinking in arriving at an expression.

(a)   3 units ⟶ 9 sticks of clay
        1 unit ⟶ 9 ÷ 3 = 3 sticks of clay

(b)   1 unit ⟶ 9 polished rocks
        3 units ⟶ 3 × 9 = 27 polished rocks

(c)   4 units ⟶ 20 bottle caps
        1 unit ⟶ 20 ÷ 4 = 5 bottle caps

(d)   2 units ⟶ $20
        1 unit ⟶ 20 ÷ 2 = $10

Students may have difficulty with (c) and (d) if they do not draw models, since in (c) they divide by 4 even though the number in the problem is 3, and in (d) they divide by 2 instead of 3, even though the problem says "3 times as many."

Exercise 8 • page 116

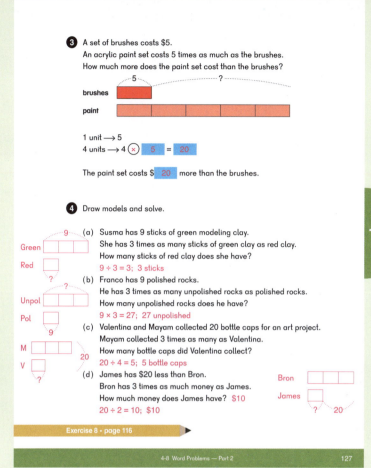

3 A set of brushes costs $5.
An acrylic paint set costs 5 times as much as the brushes.
How much more does the paint set cost than the brushes?

brushes

paint

1 unit ⟶ 5
4 units ⟶ 4 ⊗ 5 = 20

The paint set costs $ 20 more than the brushes.

4 Draw models and solve.

(a) Susma has 9 sticks of green modeling clay.
     She has 3 times as many sticks of green clay as red clay.
     How many sticks of red clay does she have?
     9 ÷ 3 = 3; 3 sticks

(b) Franco has 9 polished rocks.
     He has 3 times as many unpolished rocks as polished rocks.
     How many unpolished rocks does he have?
     9 × 3 = 27; 27 unpolished

(c) Valentina and Mayam collected 20 bottle caps for an art project.
     Mayam collected 3 times as many as Valentina.
     How many bottle caps did Valentina collect?
     20 ÷ 4 = 5; 5 bottle caps

(d) James has $20 less than Bron.
     Bron has 3 times as much money as James.
     How much money does James have? $10
     20 ÷ 2 = 10; $10

Exercise 8 • page 116

Lesson 9 2-Step Word Problems

Objective

- Solve two-step word problems involving all four operations.

Lesson Materials

- Strips of paper

Think

Pose the problem in **Think** and have students draw models. Discuss the students' models.

Ask students:

- How is the problem similar to the ones we did yesterday? (It is about multiplying and dividing. I think I can draw a comparison model.)
- How is it different? (There are more steps in this one. It's longer.)

Learn

Have students discuss the bar model in **Learn** and compare their own models with the one in the textbook.

What information do we know?

- Mei has some ribbon and she cut off 2 pieces.
- The 2nd piece of ribbon is longer than the 1st piece.
- There is still ribbon on the spool.

What do we need to find?

- How long each piece of ribbon is.

Have students begin by drawing models to represent the 1st and 2nd pieces of ribbon, adding the third bar to represent the amount of ribbon left on the spool.

Students begin by subtracting the amount of ribbon left on the spool:

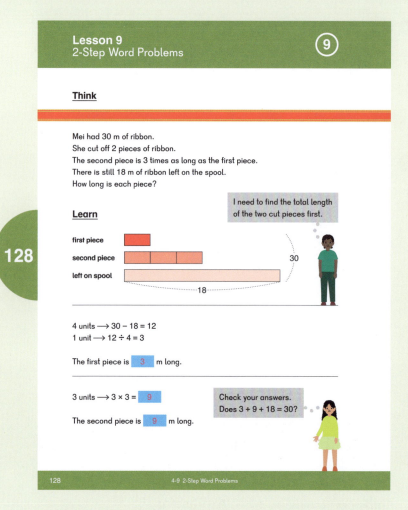

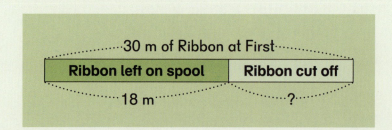

30 m − 18 m = 12 m

Then finding the length of the 2 pieces:

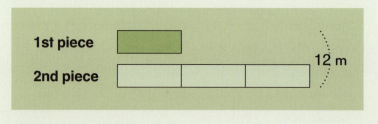

4 units ⟶ 12 m
1 unit ⟶ 12 ÷ 4 = 3 m
3 units ⟶ 3 × 3 = 9 m

Do

② This model shows a part-whole representation of addition and multiplication. Ask students:

- Why are some of the parts equal and one is not?
- Could this be drawn with two models? (Or with a comparison model?)

③ This problem is a sum and difference bar model. This pattern will be used throughout the **Dimensions Math®** program.

Sofia's thought provides a hint. If there are equal units, this becomes an easier problem. Dion can make 2 equal units by taking 5 from the total amount of dinosaurs.

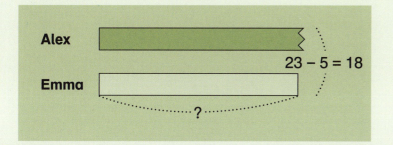

To demonstrate Sofia's thoughts, use two strips of paper that are proportional in length to the bars in the textbook. Fold behind or tear off the piece of Alex's bar that represent 5 dinosaurs to show that what remain of Alex's bar and Emma's bar are equal units.

Once we find two equal units, we can divide to find the value of one unit.

④ — ⑤ When discussing these problems, ask students:

- Why are there two models?
- Are these comparison problems?

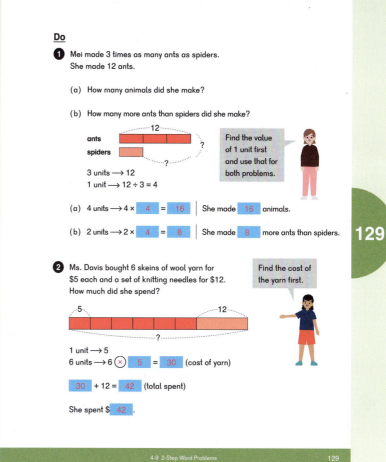

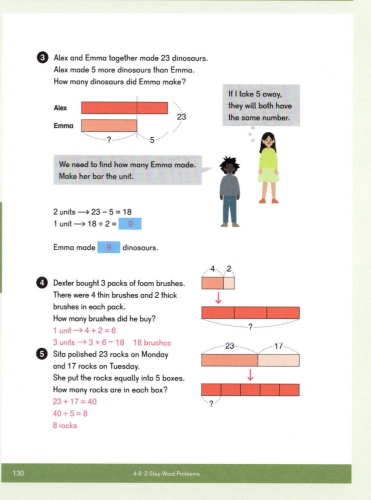

6 Discuss Alex's thought. Ask students:

• Why is Dion being used to represent 1 unit?

Additionally, this problem can be solved similarly to **3** by subtracting the difference between Mei and Dion, leaving 2 equal units.

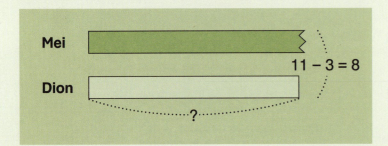

2 units ⟶ 8 turtles
1 unit ⟶ 8 ÷ 2 = 4 turtles

"Mei made 4 + 3, or 7, turtles."

7 Students may need to draw the two steps separately to see "3 times as many," and then find how many flowers in all.

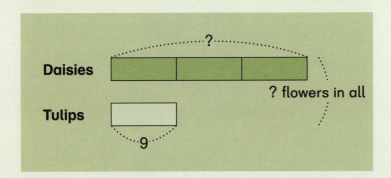

1 unit ⟶ 9 flowers
4 units ⟶ 4 × 9 = 36 flowers

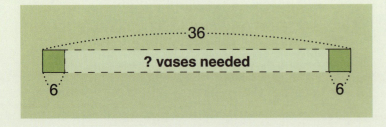

36 ÷ 6 = 6. Asimah uses 6 vases.

8 — **9** Discuss the models students draw and any alternative methods they use to solve the problems.

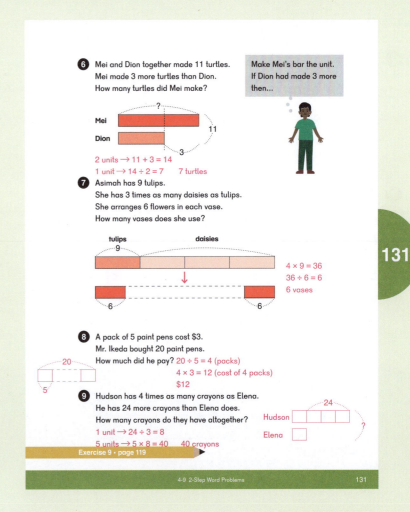

Exercise 9 • page 119

Lesson 10 Practice

Objective

- Practice topics from the chapter.

Have students practice with activities from the chapter to ensure they know their multiplication and division facts for 2 through 5.

Provide additional support and practice opportunities as needed.

5 A sample model is given. Students may draw the models differently to answer individual problems.

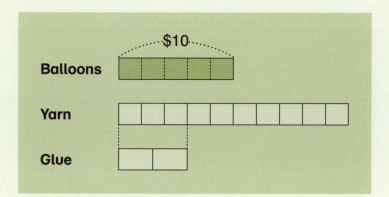

(a)   5 units $\longrightarrow$ $10
      1 unit $\longrightarrow$ $10 \div 5 = $2

(b)   1 unit $\longrightarrow$ $2
      10 units $\longrightarrow$ $10 \times $2 = $20

(c)   1 unit $\longrightarrow$ $2
      3 units $\longrightarrow$ $3 \times $2 = $6
            (3 units of yarn = 2 units of glue)
      1 unit of glue $\longrightarrow$ $6 \div 2 = $3

(e)
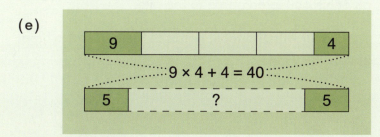

(f)
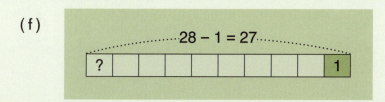

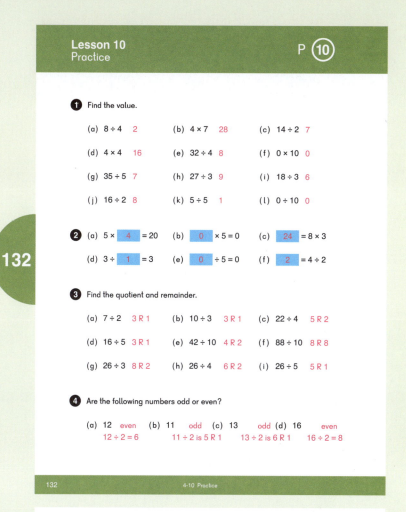

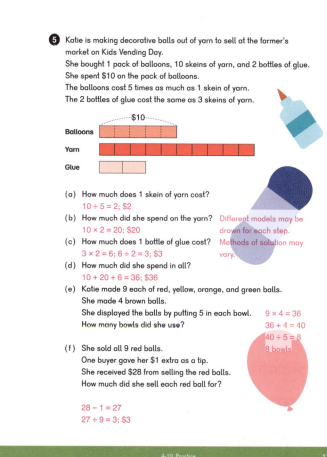

6 (a)

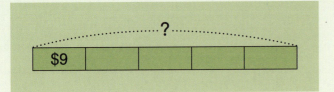

1 unit ⟶ $9

5 units ⟶ 5 × $9 = $45

Josef received $45.

(b)

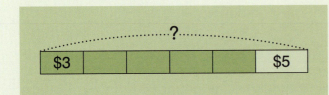

1 unit ⟶ $3

5 units ⟶ 5 × $3 = $15

$15 + $5 = $20

Josef spent $20.

(c)

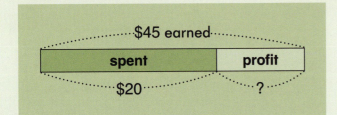

7 Arman represents the unit. Evan has 1 unit + 2 more pinecones. Mila has twice as many as Evan, or 2 units + 4 more pinecones. Altogether they have 4 units + 6 more pinecones = 30.

4 units ⟶ 30 − 6 = 24

1 unit ⟶ 24 ÷ 4 = 6 pinecones

8 Encourage students to consider **7** to help draw a model.

Activity

▲ **Multiplication and Division Kaboom**

Materials: Kaboom Cards (BLM), multiplication and division fact cards for 0 to 5

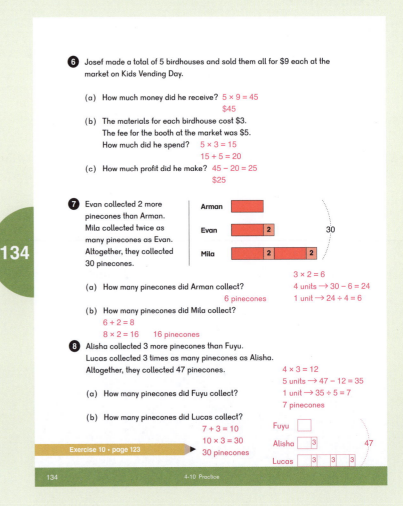

Shuffle fact cards and 3 Kaboom Cards (BLM) together and place them facedown in a pile. Players take turns drawing a card and stating the product or quotient.

Students keep the cards they answer correctly, and return the ones that they answer incorrectly. When a student draws a Kaboom Card (BLM), he must return all of his collected cards to the pile.

The player with the most cards at the end of the time limit is the winner.

Exercise 10 • page 123

Objective

- Review content from Chapters 1 through 4.

Use **Review 1** to practice and reinforce content and skills from the first four chapters.

---

1 In the number 2,846...

(a) Which digit is in the thousands place? 2
(b) What is the value of the digit 8? 800
(c) In what place is the digit 4? tens
(d) What number is 1,000 less than this number? $2,846 - 1000 = 1,846$
(e) What number is 900 more than this number? $2,846 + 900 = 3,746$
(f) What number is 6 tens less than this number? $2,846 - 60 = 2,786$

2

4,189	3,899	4,086	3,914

(a) In which of these numbers is the digit in the tens place 1 more than the digit in the hundreds place? 3,899
(b) Arrange the numbers in order from least to greatest.
(c) Find the sum of the least and greatest of these numbers.
(d) Find the difference between the least and greatest of these numbers.

(b) 3,899, 3,914, 4,086, 4,189　　(c) $3,899 + 4,189 = 8,088$　　(d) $4,189 - 3,899 = 290$

3 Round 4,549...

(a) To the nearest ten. 4,550
(b) To the nearest hundred. 4,500
(c) To the nearest thousand. 5,000

135

---

4 What sign, >, <, or =, goes in the ◯?

(a) $5,000 + 500 + 85$ ⊙> $5,000 + 5 + 85$

(b) $8 + 40 + 600 + 3,000$ ⊙< $2,000 + 6,000 + 400 + 8$

(c) 210 tens ⊙= 21 hundreds

(d) 3 thousands + 65 hundreds ⊙= 9 thousands + 50 tens

5 (a) $38 + 47 = \boxed{90} - 5$　　　　(b) $82 - 25 = \boxed{50} + 7$

(c) $740 - 90 = \boxed{640} + 10$　　　(d) $475 + 98 = \boxed{575} - 2$

(e) $60 + \boxed{8} = 100 - 32$　　　　(f) $2,000 - 480 = 1,000 + \boxed{520}$

(g) $6 \times 3 = 15 + \boxed{3}$　　　　(h) $4 \times 9 = 40 - \boxed{4}$

(i) $6 \times 6 = \boxed{9} \times 4$　　　　(j) $2 \times 4 = 16 \div \boxed{2}$

(k) $6 \times 0 = \boxed{0} \div 6$　　　　(l) $26 = 4 \times \boxed{6} + 2$

6 Estimate, then find the value. Estimations may vary.

(a) $3,875 + 283$　　　　　(b) $7,004 - 2,207$
about 4,200; 4,150　　　　about 5,000; 4,797
(c) $6,145 - 724$　　　　　(d) $2,345 + 2,655$
about 5,400; 5,421　　　　about 5,000; 5,000

136

　　　　Teacher's Guide 3A Chapter 4　　　　157

Brain Works

★ Secret Math Messages

Materials: Secret Math Message Decoder (BLM)

Students fill in the Secret Math Message Decoder (BLM) with the products to math equations. They then create blank lines for a secret message, and write an expression under each blank line so the message can be decoded.

Students trade decoder sheets with classmates and try to decode each other's message. Example:

Secret message:

___	___	___	___
4×8	3×7	9×4	6×3

___	___	___	___	___ !
$16 \div 4$	0×4	$4 \div 4$	6×5	$16 \div 2$

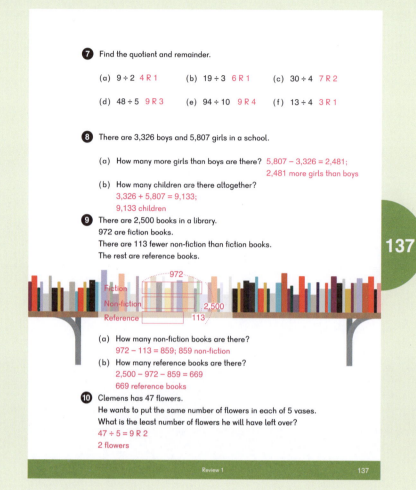

7. Find the quotient and remainder.

(a) $9 \div 2$ **4 R 1** (b) $19 \div 3$ **6 R 1** (c) $30 \div 4$ **7 R 2**

(d) $48 \div 5$ **9 R 3** (e) $94 \div 10$ **9 R 4** (f) $13 \div 4$ **3 R 1**

8. There are 3,326 boys and 5,807 girls in a school.

(a) How many more girls than boys are there? **5,807 − 3,326 = 2,481;**
    2,481 more girls than boys

(b) How many children are there altogether?
    3,326 + 5,807 = 9,133;
    9,133 children

9. There are 2,500 books in a library.
   972 are fiction books.
   There are 113 fewer non-fiction than fiction books.
   The rest are reference books.

(a) How many non-fiction books are there?
    972 − 113 = 859; 859 non-fiction

(b) How many reference books are there?
    2,500 − 972 − 859 = 669
    669 reference books

10. Clemens has 47 flowers.
    He wants to put the same number of flowers in each of 5 vases.
    What is the least number of flowers he will have left over?
    47 ÷ 5 = 9 R 2
    2 flowers

Review 1   137

11. Xavier's dog weighs 36 lb.
    His dog weighs 4 times as much as his cat.
    What is the total weight of both animals?
    1 unit → 36 ÷ 4 = 9
    5 units → 5 × 9 = 45
    45 lb

12. Diego is making airplanes.
    Each airplane uses 3 craft sticks and 1 clothespin.
    Diego has 28 craft sticks and 12 clothespins.
    He makes as many airplanes as he can.
    How many airplanes did he make?
    28 ÷ 3 is 9 R 1; 9 airplanes

13. Fatima had 3 times as many bracelets as Hannah.
    Then, Fatima lost 3 bracelets.
    The girls now have 17 bracelets altogether.
    How many bracelets does Hannah have? **4 unit → 3 = 20**
    1 unit → 20 ÷ 4 = 5
    5 bracelets

14. Allison bought 24 sticks of modeling clay.
    She put them equally into 4 bags. **24 ÷ 4 = 6**
    She used the clay in 2 of the bags to make something. **6 × 2 = 12**
    How many sticks of clay does she have left? **24 − 12 = 12**
    12 sticks

15. How many different ways can 12 children
    be divided equally into more than 1 team? **12 ÷ 6 = 2   6 teams of 2 children**
    Each team must have at least 2 children. **12 ÷ 4 = 3   4 teams of 3 children**
    12 ÷ 3 = 4   3 teams of 4 children
    12 ÷ 2 = 6   2 teams of 6 children
    4 different ways

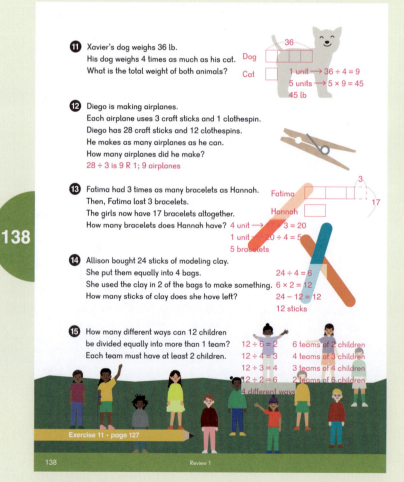

Exercise 11 • page 127

138   Review 1

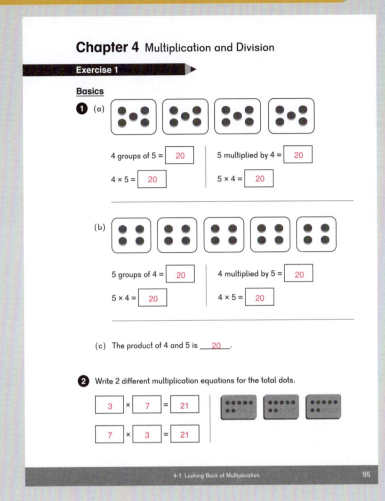

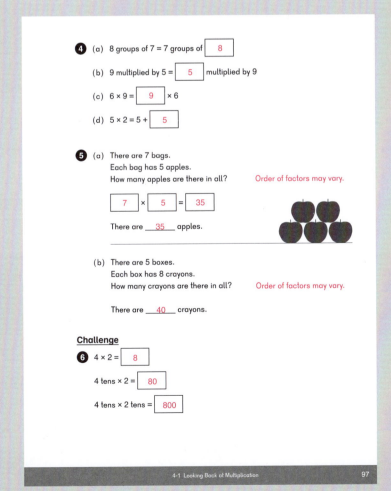

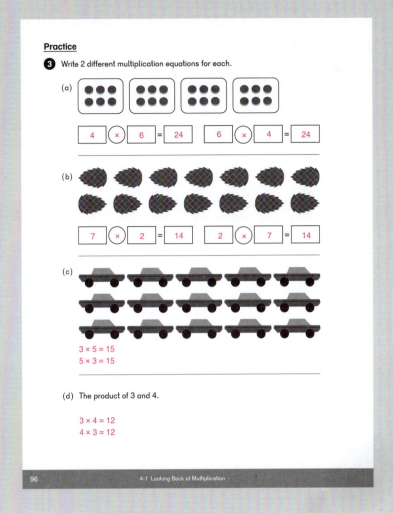

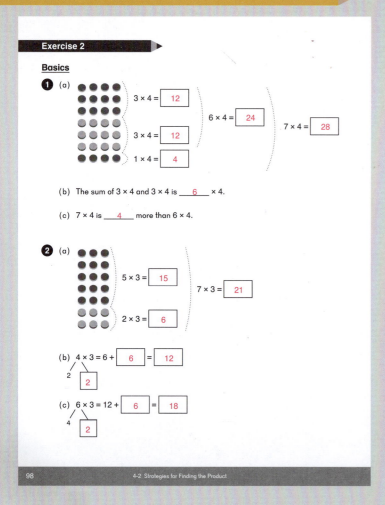

Exercise 2

Basics

1 (a)

$3 \times 4 = \boxed{12}$

$3 \times 4 = \boxed{12}$ $6 \times 4 = \boxed{24}$ $7 \times 4 = \boxed{28}$

$1 \times 4 = \boxed{4}$

(b) The sum of 3×4 and 3×4 is $\underline{\ 6\ } \times 4$.

(c) 7×4 is $\underline{\ 4\ }$ more than 6×4.

2 (a)

$5 \times 3 = \boxed{15}$ $7 \times 3 = \boxed{21}$

$2 \times 3 = \boxed{6}$

(b) $4 \times 3 = 6 + \boxed{6} = \boxed{12}$

    2   $\boxed{2}$

(c) $6 \times 3 = 12 + \boxed{6} = \boxed{18}$

    4   $\boxed{2}$

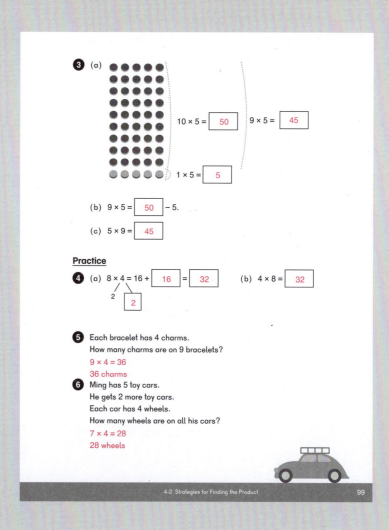

3 (a)

$10 \times 5 = \boxed{50}$ $9 \times 5 = \boxed{45}$

$1 \times 5 = \boxed{5}$

(b) $9 \times 5 = \boxed{50} - 5$.

(c) $5 \times 9 = \boxed{45}$

Practice

4 (a) $8 \times 4 = 16 + \boxed{16} = \boxed{32}$ (b) $4 \times 8 = \boxed{32}$

    2   $\boxed{2}$

5 Each bracelet has 4 charms.
How many charms are on 9 bracelets?
$9 \times 4 = 36$
36 charms

6 Ming has 5 toy cars.
He gets 2 more toy cars.
Each car has 4 wheels.
How many wheels are on all his cars?
$7 \times 4 = 28$
28 wheels

7 Complete the multiplication tables.

×	1	3	5	4	10	2
5	5	15	25	20	50	10
2	2	6	10	8	20	4
7	7	21	35	28	70	14

×	5	10	3	2	4	1
2	10	20	6	4	8	2
4	20	40	12	8	16	4
8	40	80	24	16	32	8
3	15	30	9	6	12	3
6	30	60	18	12	24	6
9	45	90	27	18	36	9
5	25	50	15	10	20	5
10	50	100	30	20	40	10

×	1	3	5	4	10	2
10	10	30	50	40	100	20
1	1	3	5	4	10	2
9	9	27	45	36	90	18

×	5	2	4	7	3	8	1	6	10	9
5	25	10	20	35	15	40	5	30	50	45
3	15	6	12	21	9	24	3	18	30	27
4	20	8	16	28	12	32	4	24	40	36

Challenge

8 (a) $8 \times 4 = 40 - \boxed{8} = \boxed{32}$

(b) The sum of 2×50 and 3×50 is $\underline{\ 5\ } \times 50$.

(c) The sum of 56×2 and 56×2 is $56 \times \boxed{4}$.

Exercise 3

Basics

1. (a) Divide 12 counters into 4 equal groups.
   Draw a picture to show this.

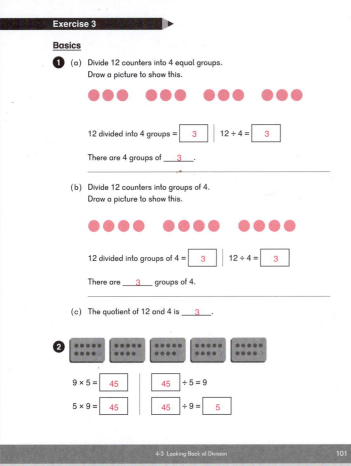

12 divided into 4 groups = 3 | 12 ÷ 4 = 3

There are 4 groups of __3__ .

(b) Divide 12 counters into groups of 4.
   Draw a picture to show this.

12 divided into groups of 4 = 3 | 12 ÷ 4 = 3

There are __3__ groups of 4.

(c) The quotient of 12 and 4 is __3__ .

2.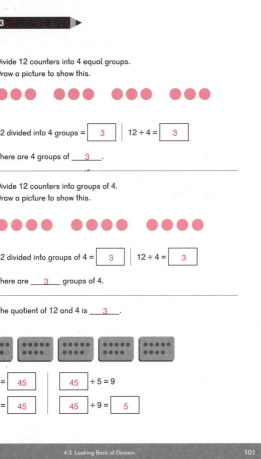

9 × 5 = 45 | 45 ÷ 5 = 9

5 × 9 = 45 | 45 ÷ 9 = 5

Practice

3. There are 18 pieces of sushi.
   Each plate can hold 3 pieces of sushi.
   Find how many plates are needed.

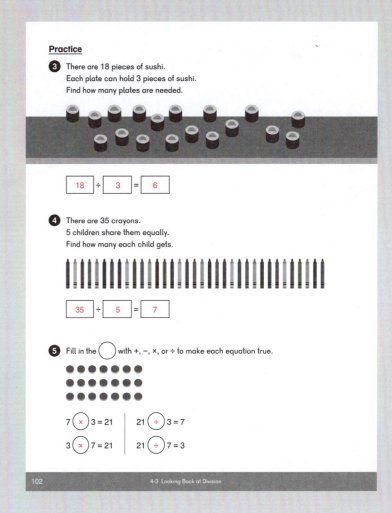

18 ÷ 3 = 6

4. There are 35 crayons.
   5 children share them equally.
   Find how many each child gets.

35 ÷ 5 = 7

5. Fill in the ◯ with +, −, ×, or ÷ to make each equation true.

7 ⊗ 3 = 21 | 21 ⊘ 3 = 7

3 ⊗ 7 = 21 | 21 ⊘ 7 = 3

6. (a) 7 × 4 = 28   (b) 5 × 5 = 25

   28 ÷ 4 = 7   25 ÷ 5 = 5

   (c) 3 × 8 = 24   (d) 4 × 9 = 36

   24 ÷ 3 = 8   36 ÷ 4 = 9

7. (a) 35 ÷ 5 = 7   (b) 18 ÷ 2 = 9

   (c) 24 ÷ 3 = 8   (d) 40 ÷ 10 = 4

8. Each person bought a sandwich for $4 and a drink for $1.
   Altogether they spent $20.
   How many people were there?
   4 + 1 = 5
   20 ÷ 5 = 4
   There were 4 people.

Challenge

9. The number in each box is the product of the two boxes below it.
   Write the missing numbers.

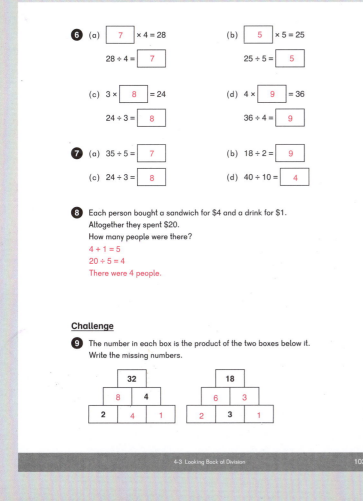

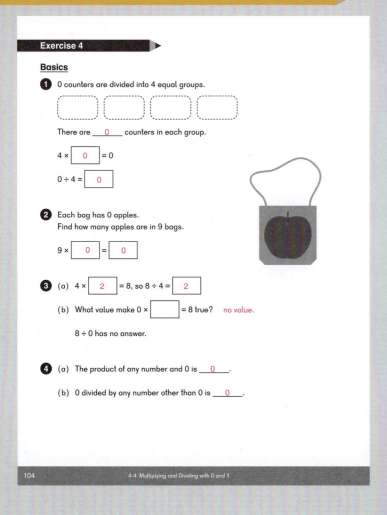

Exercise 4

Basics

1 0 counters are divided into 4 equal groups.

There are ___0___ counters in each group.

$4 \times \boxed{0} = 0$

$0 \div 4 = \boxed{0}$

2 Each bag has 0 apples.
Find how many apples are in 9 bags.

$9 \times \boxed{0} = \boxed{0}$

3 (a) $4 \times \boxed{2} = 8$, so $8 \div 4 = \boxed{2}$

(b) What value make $0 \times \boxed{} = 8$ true? no value.

$8 \div 0$ has no answer.

4 (a) The product of any number and 0 is ___0___.

(b) 0 divided by any number other than 0 is ___0___.

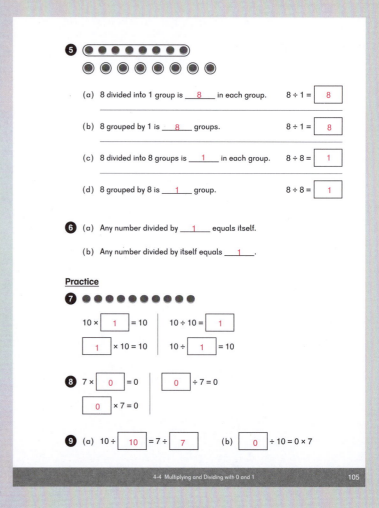

5

(a) 8 divided into 1 group is ___8___ in each group. $8 \div 1 = \boxed{8}$

(b) 8 grouped by 1 is ___8___ groups. $8 \div 1 = \boxed{8}$

(c) 8 divided into 8 groups is ___1___ in each group. $8 \div 8 = \boxed{1}$

(d) 8 grouped by 8 is ___1___ group. $8 \div 8 = \boxed{1}$

6 (a) Any number divided by ___1___ equals itself.

(b) Any number divided by itself equals ___1___.

Practice

7

$10 \times \boxed{1} = 10$ | $10 \div 10 = \boxed{1}$

$\boxed{1} \times 10 = 10$ | $10 \div \boxed{1} = 10$

8 $7 \times \boxed{0} = 0$ | $\boxed{0} \div 7 = 0$

$\boxed{0} \times 7 = 0$

9 (a) $10 \div \boxed{10} = 7 \div \boxed{7}$ (b) $\boxed{0} \div 10 = 0 \times 7$

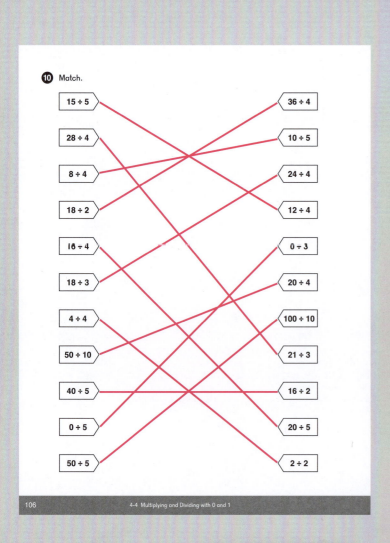

10 Match.

$15 \div 5$
$28 \div 4$
$8 \div 4$
$18 \div 2$
$16 \div 4$
$18 \div 3$
$4 \div 4$
$50 \div 10$
$40 \div 5$
$0 \div 5$
$50 \div 5$

$36 \div 4$
$10 \div 5$
$24 \div 4$
$12 \div 4$
$0 \div 3$
$20 \div 4$
$100 \div 10$
$21 \div 3$
$16 \div 2$
$20 \div 5$
$2 \div 2$

Exercise 5

Basics

1 Circle the counters to make groups of 4.

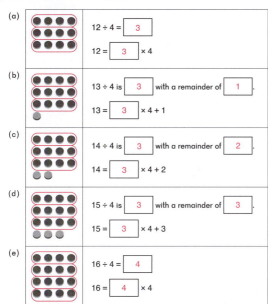

(a) $12 \div 4 = \boxed{3}$

$12 = \boxed{3} \times 4$

(b) $13 \div 4$ is $\boxed{3}$ with a remainder of $\boxed{1}$.

$13 = \boxed{3} \times 4 + 1$

(c) $14 \div 4$ is $\boxed{3}$ with a remainder of $\boxed{2}$.

$14 = \boxed{3} \times 4 + 2$

(d) $15 \div 4$ is $\boxed{3}$ with a remainder of $\boxed{3}$.

$15 = \boxed{3} \times 4 + 3$

(e) $16 \div 4 = \boxed{4}$

$16 = \boxed{4} \times 4$

(f) The remainders when dividing by 4 are __1__, __2__, and __3__.

(g) The remainders are all less than __4__.

Practice

2 There are 23 pieces of sushi.
Each plate can hold at most 3 pieces of sushi.
Find the fewest number of plates that are needed.

$\boxed{23} \div \boxed{3}$ is $\boxed{7}$ with a remainder of $\boxed{2}$.

__8__ plates are needed.

__7__ of the plates have 3 pieces of sushi.

__1__ of the plates has __2__ pieces of sushi.

3 There are 39 crayons.
5 children want to share them equally.
Find how many each child gets and how many are left over.

$\boxed{39} \div \boxed{5}$ is $\boxed{7}$ with a remainder of $\boxed{4}$.

Each child gets __7__ crayons.

__4__ crayons are left over.

4 There are 38 pennies.
If we trade in as many pennies as possible for nickels,
how many nickels and pennies will there be?

$\boxed{7} \times 5 = 35$

$\boxed{8} \times 5 = 40$

38 is between 35 and 40. $38 - 35 = \boxed{3}$

$\boxed{38} \div \boxed{5}$ is $\boxed{7}$ with a remainder of $\boxed{3}$.

There will be __7__ nickels and __3__ pennies.

5 (a) $19 \div 2$ is $\boxed{9}$ with a remainder of $\boxed{1}$.

(b) $\boxed{20} \div 3$ is 6 with a remainder of 2.

(c) $57 \div \boxed{10}$ is 5 with a remainder of 7.

(d) $33 \div \boxed{4}$ is 8 with a remainder of $\boxed{1}$.

6 Abigail has 15 flowers.
She wants to put the same number of flowers in each vase.
What different numbers of vases could she use?
1, 3, 5, or 15

7 Color the spaces where there is a remainder of 1 to help the spider find its home.

36 ÷ 5	9 ÷ 2	27 ÷ 5	10 ÷ 4	33 ÷ 10	20 ÷ 2	34 ÷ 4
12 ÷ 2	10 ÷ 3	17 ÷ 3	17 ÷ 10	19 ÷ 2	33 ÷ 4	41 ÷ 5
27 ÷ 3	13 ÷ 4	18 ÷ 2	26 ÷ 3	29 ÷ 4	14 ÷ 3	17 ÷ 4
8 ÷ 1	22 ÷ 3	41 ÷ 10	0 ÷ 4	28 ÷ 3	39 ÷ 4	6 ÷ 5
15 ÷ 10	23 ÷ 4	46 ÷ 5	0 ÷ 5	36 ÷ 5	11 ÷ 3	13 ÷ 3
10 ÷ 1	34 ÷ 5	9 ÷ 4	17 ÷ 2	31 ÷ 10	39 ÷ 5	5 ÷ 2

Challenge

8 What numbers between 24 and 49 have a remainder of 3 when divided by 5?
33, 38, 43, or 48

$5 \times 6 + 3 = 33$
$5 \times 7 + 3 = 38$
$5 \times 8 + 3 = 43$
$5 \times 9 + 3 = 48$

Exercise 6

Basics

1 A number that can be divided by 2 with no remainder is called an

___even___ number.

2 Find the quotient and remainder when each of these numbers is divided by 2.
Put a check mark in the column to show if the number is odd or even.

Number	÷ 2		Odd	Even
	Quotient	Remainder		
20	10	0		✓
19	9	1	✓	
18	9	0		✓
17	8	1	✓	
16	8	0		✓
15	7	1	✓	
14	7	0		✓
13	6	1	✓	
12	6	0		✓
11	5	1	✓	
10	5	0		✓
9	4	1	✓	
8	4	0		✓
7	3	1	✓	
6	3	0		✓
5	2	1	✓	
4	2	0		✓
3	1	1	✓	
2	1	0		✓
1	0	1	✓	
0	0	0		✓

Practice

3 Fill in the blanks with even or odd.
Students may either generalize the ideas or find the answers first.

(a) 6 + 8
even + even = even

(b) 5 + 7
odd + odd = ___even___

(c) 6 + 7
even + ___odd___ = ___odd___

(d) 5 + 8
___odd___ + ___even___ = ___odd___

(e) 12 − 4
even − even = ___even___

(f) 11 − 5
___odd___ − ___odd___ = ___even___

(g) 12 − 5
___even___ − ___odd___ = ___odd___

(h) 11 − 6
___odd___ − ___even___ = ___odd___

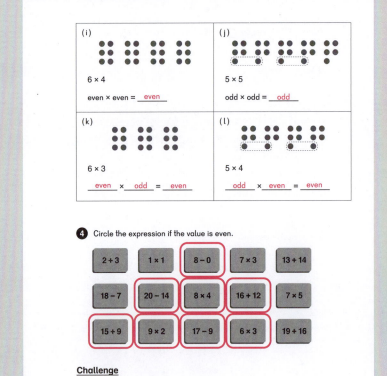

(i) 6 × 4
even × even = ___even___

(j) 5 × 5
odd × odd = ___odd___

(k) 6 × 3
___even___ × ___odd___ = ___even___

(l) 5 × 4
___odd___ × ___even___ = ___even___

4 Circle the expression if the value is even.

2 + 3	1 × 1	(8 − 0)	7 × 3	13 + 14
18 − 7	(20 − 14)	(8 × 4)	(16 + 12)	7 × 5
(15 + 9)	(9 × 2)	(17 − 9)	(6 × 3)	19 + 16

Challenge

5 Circle the expression if the value is even.

| (18 × 4) | 17 × 5 | 13 × 3 |

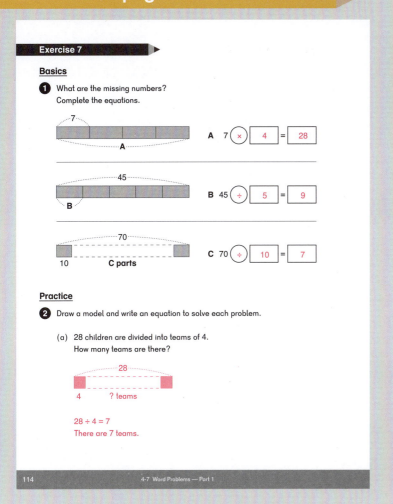

Exercise 7

Basics

1. What are the missing numbers?
Complete the equations.

A 7 ⓧ 4 = 28

B 45 ⓓ 5 = 9

C 70 ⓓ 10 = 7

10   C parts

Practice

2. Draw a model and write an equation to solve each problem.

(a) 28 children are divided into teams of 4.
How many teams are there?

28

4   ? teams

28 ÷ 4 = 7
There are 7 teams.

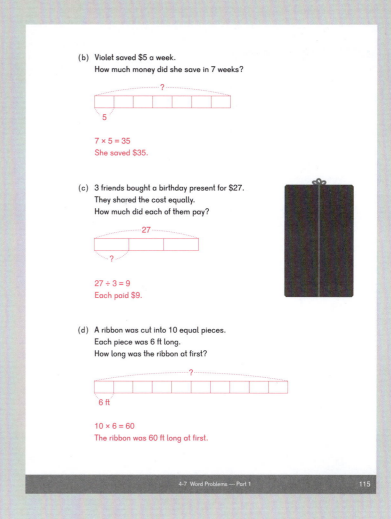

(b) Violet saved $5 a week.
How much money did she save in 7 weeks?

?

5

7 × 5 = 35
She saved $35.

(c) 3 friends bought a birthday present for $27.
They shared the cost equally.
How much did each of them pay?

27

?

27 ÷ 3 = 9
Each paid $9.

(d) A ribbon was cut into 10 equal pieces.
Each piece was 6 ft long.
How long was the ribbon at first?

?

6 ft

10 × 6 = 60
The ribbon was 60 ft long at first.

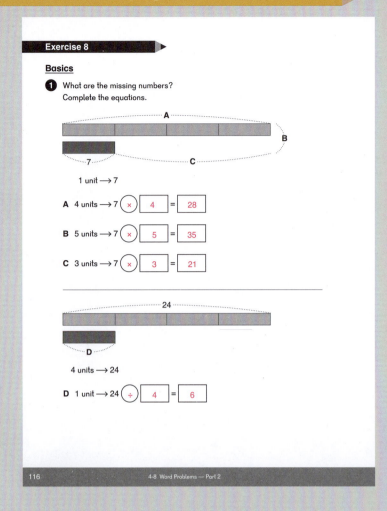

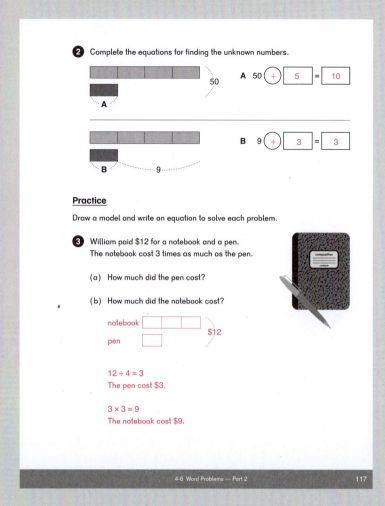

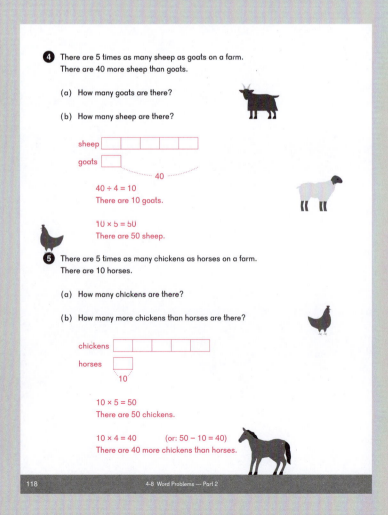

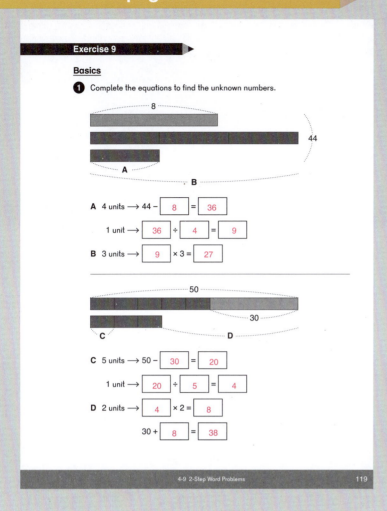

Exercise 9

Basics

1. Complete the equations to find the unknown numbers.

A 4 units → 44 − 8 = 36

1 unit → 36 ÷ 4 = 9

B 3 units → 9 × 3 = 27

C 5 units → 50 − 30 = 20

1 unit → 20 ÷ 5 = 4

D 2 units → 4 × 2 = 8

30 + 8 = 38

4-9 2-Step Word Problems   119

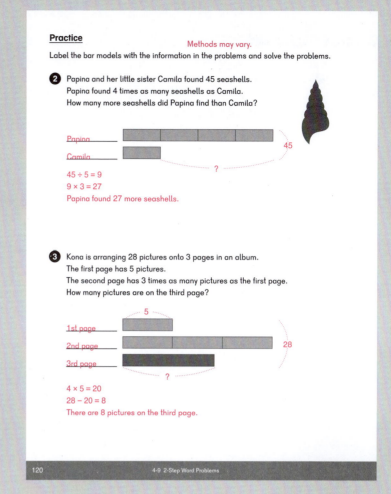

Practice   Methods may vary.

Label the bar models with the information in the problems and solve the problems.

2. Papina and her little sister Camila found 45 seashells.
Papina found 4 times as many seashells as Camila.
How many more seashells did Papina find than Camila?

Papina
Camila   45
?

45 ÷ 5 = 9
9 × 3 = 27
Papina found 27 more seashells.

3. Kona is arranging 28 pictures onto 3 pages in an album.
The first page has 5 pictures.
The second page has 3 times as many pictures as the first page.
How many pictures are on the third page?

5
1st page
2nd page   28
3rd page
?

4 × 5 = 20
28 − 20 = 8
There are 8 pictures on the third page.

120   4-9 2-Step Word Problems

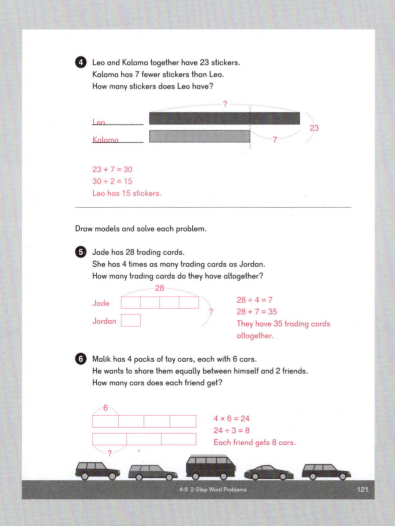

4. Leo and Kalama together have 23 stickers.
Kalama has 7 fewer stickers than Leo.
How many stickers does Leo have?

?
Leo
Kalama   23
7

23 + 7 = 30
30 ÷ 2 = 15
Leo has 15 stickers.

Draw models and solve each problem.

5. Jade has 28 trading cards.
She has 4 times as many trading cards as Jordan.
How many trading cards do they have altogether?

28
Jade
?
Jordan

28 ÷ 4 = 7
28 + 7 = 35
They have 35 trading cards altogether.

6. Malik has 4 packs of toy cars, each with 6 cars.
He wants to share them equally between himself and 2 friends.
How many cars does each friend get?

6
?

4 × 6 = 24
24 ÷ 3 = 8
Each friend gets 8 cars.

4-9 2-Step Word Problems   121

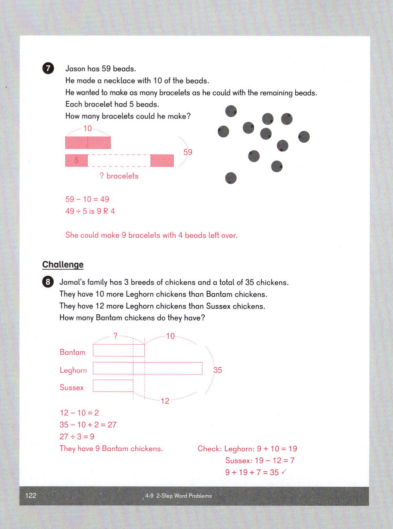

7. Jason has 59 beads.
He made a necklace with 10 of the beads.
He wanted to make as many bracelets as he could with the remaining beads.
Each bracelet had 5 beads.
How many bracelets could he make?

10
5   59
? bracelets

59 − 10 = 49
49 ÷ 5 is 9 R 4

She could make 9 bracelets with 4 beads left over.

Challenge

8. Jamal's family has 3 breeds of chickens and a total of 35 chickens.
They have 10 more Leghorn chickens than Bantam chickens.
They have 12 more Leghorn chickens than Sussex chickens.
How many Bantam chickens do they have?

?   10
Bantam
Leghorn   35
Sussex
12

12 − 10 = 2
35 − 10 + 2 = 27
27 ÷ 3 = 9
They have 9 Bantam chickens.   Check: Leghorn: 9 + 10 = 19
Sussex: 19 − 12 = 7
9 + 19 + 7 = 35 ✓

122   4-9 2-Step Word Problems

Exercise 10

Check

1. (a) Find the product of 3 and 7.
   21

   (b) Find the quotient of 0 divided by 8.
   0

   (c) Find the quotient and remainder for 35 divided by 4.
   8 R 3

2. (a) 26 = 8 × [3] + 2

   (b) 5 + 5 + 5 + 5 + [15] = 7 × 5

   (c) 7 × 4 = 5 × 4 + [8]

   (d) 9 × 4 = 10 × 4 − [4]

3. Tomas has an odd number of flowers.
   He puts the same number of flowers into 4 vases.
   How many flowers could he have left over?
   1 or 3

4. Multiply or divide.

2 × 7 = 14 **S**	40 ÷ 4 = 10 **O**	36 ÷ 4 = 9 **E**
3 × 7 = 21 **L**	5 × 5 = 25 **R**	18 ÷ 3 = 6 **I**
8 ÷ 8 = 1 **T**	4 × 3 = 12 **P**	4 × 4 = 16 **N**
4 × 6 = 24 **F**	8 ÷ 4 = 2 **W**	5 × 9 = 45 **S**
3 × 5 = 15 **I**	80 ÷ 10 = 8 **K**	20 ÷ 5 = 4 **C**
25 ÷ 5 = 5 **M**	2 × 0 = 0 **U**	28 ÷ 4 = 7 **E**
3 × 6 = 18 **D**	5 × 4 = 20 **H**	9 ÷ 3 = 3 **A**

Write the letters that match the answers above to learn something weird but true.

K	E	T	C	H	U	P		W	A	S
17	8	7	1	4	20	0	12	13	2	3	14	11

F	I	R	S	T			S	O	L	D		A	S
24	6	25	45	1		19	14	10	21	18	29	3	45

		M	E	D	I	C	I	N	E			
27	26	5	9	18	15	4	6	16	7	22	30	23

5. 5 children share 38 crayons equally.
   How many crayons are left over?
   38 ÷ 5 is 7 R 3
   3 crayons are left over.

6. A pack of 10 notepads costs $5.
   Jasmine has $42.

   (a) How many packs of notepads could she buy?

   (b) If she buys 5 packs of notepads, how much money will she have left over?

   (c) If she spends $35 on the packs of notepads, how many notepads will she have?
   42 ÷ 5 is 8 R 2
   She could buy 8 packs of notepads.

   5 × 5 = 25   35 ÷ 5 = 7
   42 − 25 = 17   7 × 10 = 70
   She will have $17 left over.   She will have 70 notepads.

7. Mia has 3 times as many ribbons as Asima.
   Altogether they have 20 ribbons.
   How many more ribbons does Mia have than Asima?

   Mia [ ][ ][ ]
   Kaitlyn [ ] 20
            ?

   20 ÷ 2 = 10
   Mia has 10 more ribbons than Kaitlyn.

Challenge

8. Complete the cross-number puzzles.

(a)
7	×	4	=	28
×		×		+
10	×	3	=	30
=		=		=
70	−	12	=	58

(b)
5	×	8	=	40
×		×		−
2	×	3	=	6
=		=		=
10	+	24	=	34

9. Logan has twice as many comic books as Jade.
   After Jade buys 9 more comic books, she has twice as many comic books as Logan.
   How many comic books did Jade have at first?

   Before
   Jade [ ]
   Logan [ ][ ]

   9 ÷ 3 = 3

   Jade had 3 comic books at first.

   After
   Jade [ ][ ][ ] 9
   Logan [ ][ ]

Exercise 11

Check

1 What is the value of each digit?

(a) 9,208

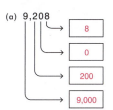

→	8
→	0
→	200
→	9,000

(b) 3,190

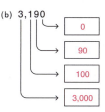

→	0
→	90
→	100
→	3,000

(c) In 9,208, the digit 9 is in the ___thousands___ place.

(d) In 3,190, the digit 9 is in the __tens__ place.

(e) 3,190 is the same as __319__ tens.

(f) Write 9,208 in words.

nine thousand, two hundred eight

2 Write the greatest and least 4-digit number you can make using all the digits.

Digits	Greatest	Least
9, 0, 8, 0	9,800	8,009
2, 2, 1, 4	4,221	1,224

---

3 Write > or < in the ○.

6,043 (<) 3 + 400 + 6,000

44 hundreds (<) 4,000 + 40 + 600

8,450 (>) 32 hundreds 5 thousands

4 Draw an arrow to show the location of 1,575 on the number line.

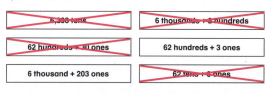

1,500   ↑   1,600

5 Round 4,845…

(a) To the nearest thousand. | 5,000 |

(b) To the nearest hundred. | 4,800 |

(c) To the nearest ten. | 4,850 |

6 Cross out the incorrect answers.

6,203 is the same as…

~~6,203 tens~~	6 thousands + 2 hundreds
~~62 hundreds + 40 ones~~	62 hundreds + 3 ones
6 thousand + 203 ones	~~62 tens + 3 ones~~

---

7 Use mental calculation to find the value.

(a) 580 + 250 = | 830 | (b) 810 – 740 = | 70 |

(c) 477 + 199 = | 676 | (d) 462 – 197 = | 265 |

(e) 5,000 – 260 = | 4,740 | (f) 600 – 344 = | 256 |

8 (a) Is 4,239 + 3,440 closer to 7,000 or 8,000? __8,000__

(b) Is 4,239 – 3,440 closer to 700 or 800? __800__

9 (a) 7 × 3 = | 3 | × 7   (b) 6 × 4 = | 3 | × 8

(c) 27 ÷ 3 = | 3 | × 3   (d) 38 = | 9 | × 4 + 2

(e) 17 = 8 × | 2 | + 1   (f) 10 × 3 = | 5 | × 5 + 5

10 What number do you need to add to the sum of 2,420 and 3,980 to get the number 8,000?

2,420 + 3,980 = 6,400
8,000 – 6,400 = 1,600
You need to add 1,600.

---

11 Nora saved $2,000 to spend on a computer. She bought a gaming laptop for $1,349.

(a) How much money does she have left?

(b) A business laptop costs $458 more than the gaming laptop. How much does the business laptop cost?

(c) A tablet computer costs $651 less than the gaming laptop. How much does the tablet computer cost?

2,000 – 1,349 = 651
She has $651 left.

1,349 + 458 = 1,807
It costs $1,807.

1,349 – 651 = 698
The tablet computer costs $698.

12 There are 1,240 lemon and orange trees in an orchard. If there are 590 lemon trees, how many more orange trees than lemon trees are there?

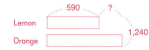

Lemon [590   ] ?
Orange [   ] 1,240

1,240 – 590 = 650
650 – 590 = 60
There are 60 more orange trees than lemon trees.

13 The eggs from some chickens were packed into 5 egg cartons.
Each egg carton holds 6 eggs.
10 eggs broke before they were packed.
If each chicken laid 4 eggs that day, how many chickens laid eggs?

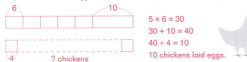

5 × 6 = 30
30 + 10 = 40
40 ÷ 4 = 10
10 chickens laid eggs.

14 There are 30 red, yellow, and pink rose bushes in a garden.
There are twice as many red rose bushes as yellow rose bushes.
There are 6 pink rose bushes.
How many yellow rose bushes are there?

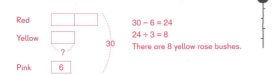

30 − 6 = 24
24 ÷ 3 = 8
There are 8 yellow rose bushes.

15 List the numbers less than 30 that have a remainder of 2 when divided by 3 or 4.
When divided by 3: 29, 26, 23, 20, 17, 14, 11, 8, 5
When divided by 4: 26, 22, 18, 14, 10, 6
Both: 26, 14

Or: Only even numbers have a remainder of 2 when divided by 4.
Check even numbers less than 30 that are not products of 4:
6, 10, 14, 18, 22, 26,
6 and 18 will have no remainders when divided by 3.
14 and 26 have a remainder of 2 when divided by 3.

Challenge

16 Write +, −, ×, or ÷ in the ◯.

(a) 8 ⊕ 12 = 5 ⊗ 4

(b) 12 ⊘ 2 = 24 ⊘ 4

(c) 24 ⊘ 4 = 12 ⊖ 6

(d) 27 ⊕ 3 = 20 ⊕ 10

(e) 8 × 4 ⊖ 8 = 6 ⊗ 4

(f) 7 × 4 ⊖ 8 = 10 ⊗ 2

17 Complete the cross number puzzle.

(a)

12	÷	3	=	4
÷		+		÷
2	×	2	=	4
=		=		=
6	−	5	=	1

(b)

32	÷	4	=	8
−		+		+
20	÷	2	=	10
=		=		=
12	+	6	=	18

18 To find the difference between 715 and 285, Mei first mentally added 15 to each number and then calculated 730 − 300 = 430.
Does her method work?
Explain.

	715
	285

Adding the same number to two numbers does not change the difference between those two numbers.

Suggested number of class periods: 9–10

	Lesson	Page	Resources		Objectives
	Chapter Opener	p. 175	TB:	p. 139	Investigate multi-digit multiplication.
1	Multiplying Ones, Tens, and Hundreds	p. 176	TB: WB:	p. 140 p. 133	Multiply a multiple of 10 by a one-digit number.
2	Multiplication Without Regrouping	p. 179	TB: WB:	p. 144 p. 135	Multiply a two-digit number by a one-digit number without regrouping.
3	Multiplication with Regrouping Tens	p. 182	TB: WB:	p. 148 p. 137	Multiply a two-digit number by a one-digit number with regrouping tens.
4	Multiplication with Regrouping Ones	p. 185	TB: WB:	p. 152 p. 139	Multiply a two-digit number by a one-digit number with regrouping ones.
5	Multiplication with Regrouping Ones and Tens	p. 188	TB: WB:	p. 156 p. 142	Multiply a two-digit number by a one-digit number with regrouping tens and ones.
6	Practice A	p. 190	TB: WB:	p. 160 p. 145	Practice multiplying ones, tens, and hundreds by a one-digit number.
7	Multiplying a 3-Digit Number with Regrouping Once	p. 191	TB: WB:	p. 162 p. 148	Multiply a three-digit number with regrouping tens or ones.
8	Multiplication with Regrouping More Than Once	p. 193	TB: WB:	p. 166 p. 151	Multiply a three-digit number with regrouping in more than one place.
9	Practice B	p. 196	TB: WB:	p. 170 p. 154	Practice multiplying three-digit numbers.
	Workbook Solutions	p. 197			

This chapter introduces the standard multiplication algorithm, in which numbers are written in a vertical format with the digits aligned.

In this chapter, students will multiply two and three-digit numbers by 2, 3, 4, and 5. They should be fluent with multiplication facts for 2 to 5. This chapter precedes multiplication and division by 6, 7, 8, and 9 so students can focus on the algorithm using facts they should know well by now.

Students will use place-value discs to enrich their conceptual understanding of the algorithm as they gain procedural fluency. Students should use place-value discs when first solving the **Think** problems.

Lessons will start by showing each place being multiplied separately, with students thinking of the number split into hundreds, tens, and ones. Students will be able to solve some problems mentally using this approach.

$$215 \times 3 = 600 + 30 + 15$$
$$\overset{\diagup\ |\ \diagdown}{200\ \ 10\ \ 5}$$

Students will also be shown a partial product method:

$$
\begin{array}{r}
7\,5\,5 \\
\times \quad\ \ 3 \\
\hline
1\,5 \quad \longleftarrow\ 5 \times 3 \\
1\,5\,0 \quad \longleftarrow\ 50 \times 3 \\
\underline{2{,}1\,0\,0} \quad \longleftarrow\ 700 \times 3 \\
2{,}2\,6\,5
\end{array}
$$

Students can use this approach until they feel more confident with the standard algorithm, which is a more concise method. In the standard algorithm, regrouped digits are added immediately, rather than adding only at the end.

Standard Algorithm

Below is a general procedure for demonstrating the multiplication algorithm with place-value discs.

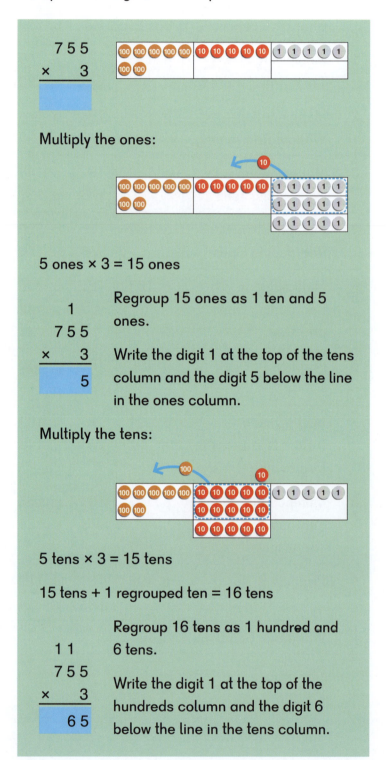

Multiply the ones:

5 ones × 3 = 15 ones

Regroup 15 ones as 1 ten and 5 ones.

Write the digit 1 at the top of the tens column and the digit 5 below the line in the ones column.

Multiply the tens:

5 tens × 3 = 15 tens

15 tens + 1 regrouped ten = 16 tens

Regroup 16 tens as 1 hundred and 6 tens.

Write the digit 1 at the top of the hundreds column and the digit 6 below the line in the tens column.

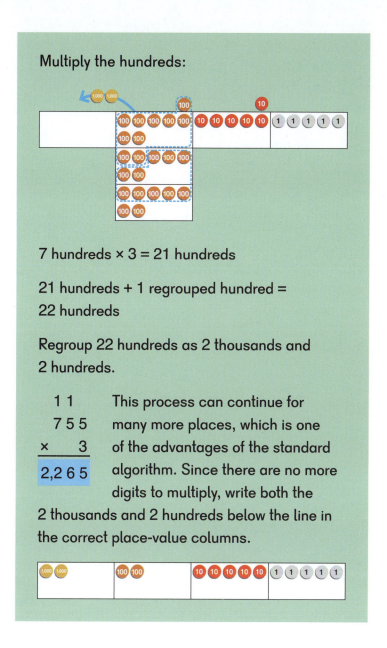

Multiply the hundreds:

7 hundreds × 3 = 21 hundreds

21 hundreds + 1 regrouped hundred = 22 hundreds

Regroup 22 hundreds as 2 thousands and 2 hundreds.

```
  1 1
  7 5 5
×     3
―――――――
2,2 6 5
```

This process can continue for many more places, which is one of the advantages of the standard algorithm. Since there are no more digits to multiply, write both the 2 thousands and 2 hundreds below the line in the correct place-value columns.

- Model the **Learn** procedure, and have students work the problem with place-value discs while it's being modeled.

- Open the textbook and discuss the **Learn** section, making sure students understand how to relate the steps they have done with place-value discs to the written algorithm.

- Struggling students may work the **Do** problems using the place-value discs **first**, then compare them to the textbooks.

- Have students work in pairs with the place-value discs, whiteboards, and textbooks.

- As students gain confidence, let them copy any remaining problems from the textbook into notebooks or whiteboards and work them without place-value discs. Students could also draw place-value discs as an interim step.

The goal is to have the students understand how the algorithm works and do it without using place-value discs.

Note on managing manipulatives

Typically, students can solve **Do** problems pictorially.

Due to the amount of manipulatives used in this chapter, some suggestions on structuring the lesson are included below:

- Pose the **Think** problem and allow adequate time for students to work out solutions. After they solve the problem, have students share and discuss their methods. The textbook does not need to be open at this time.

Materials

- Dry erase markers
- Dry erase sleeves
- Paper clips
- Pencils
- Place-value discs
- Playing cards
- Two-color counters
- Whiteboards

Blackline Masters

- Greatest Product – 2-Digit Game Board
- Greatest Product – 3-Digit Game Board
- How Many Can You Find? × 2
- How Many Can You Find? × 3
- Multiplication Symbol Cards
- Number Cards
- Spin and Multiply Recording Sheet
- Tens and Hundreds Spinner

Activities

Activities included in this chapter are designed to provide practice with multiplying up to three-digit numbers. They can be used after students complete the **Do** questions, or any time additional practice is needed.

Chapter Opener

Objective

- Investigate multi-digit multiplication.

Lesson Materials

- Two-color counters

Discuss the examples in the **Chapter Opener**. Students should note that they can solve 4 × 12 in different ways based on what they have already learned about multiplication.

Provide pairs of students with two-color counters and ask them to think of other ways they could find the product of 4 and 12, then challenge them to use these methods to find 13 × 5.

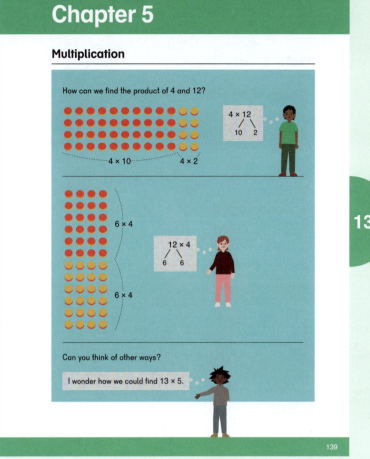

Lesson 1 Multiplying Ones, Tens, and Hundreds

Objective

- Multiply a multiple of 10 by a one-digit number.

Lesson Materials

- Place-value discs

Think

Provide students with place-value discs and have them work the **Think** problem independently.

Have students write expressions to find the total for each item and share how they found their answers.

Learn

Work through the **Think** problem with the students as demonstrated in **Learn**. Have students work along with place-value discs as the steps are modeled.

Help students see that they can view 50 × 3 as 5 tens × 3. Just as 5 apples × 3 are 15 apples and 5 sheep × 3 are 15 sheep, 5 tens × 3 are 15 tens, and 5 hundreds × 3 are 15 hundreds.

After the students have completed working the problem with place-value discs, have them compare their methods from **Think** with the method shown in the textbook.

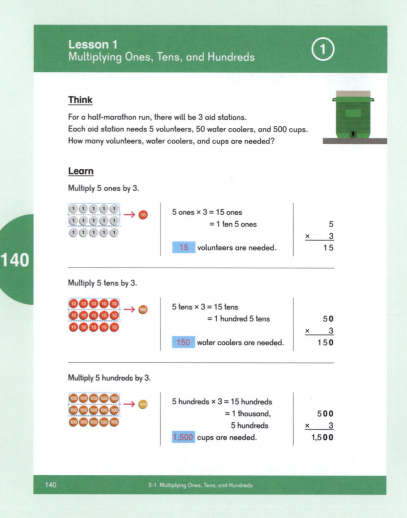

Lesson 1
Multiplying Ones, Tens, and Hundreds ①

Think

For a half-marathon run, there will be 3 aid stations.
Each aid station needs 5 volunteers, 50 water coolers, and 500 cups.
How many volunteers, water coolers, and cups are needed?

Learn

Multiply 5 ones by 3.

5 ones × 3 = 15 ones
= 1 ten 5 ones

$$\begin{array}{r} 5 \\ \times\ 3 \\ \hline 15 \end{array}$$

15 volunteers are needed.

Multiply 5 tens by 3.

5 tens × 3 = 15 tens
= 1 hundred 5 tens

$$\begin{array}{r} 50 \\ \times\ 3 \\ \hline 150 \end{array}$$

150 water coolers are needed.

Multiply 5 hundreds by 3.

5 hundreds × 3 = 15 hundreds
= 1 thousand,
5 hundreds

$$\begin{array}{r} 500 \\ \times\ 3 \\ \hline 1,500 \end{array}$$

1,500 cups are needed.

140   5-1 Multiplying Ones, Tens, and Hundreds

Do

Have students show the numbers with place-value discs if needed. Most should be working with just the pages in the textbook.

Have students rewrite the equations vertically. Check that students are aligning the digits correctly.

2 Ensure students understand the regrouping process. For example, in (a), 24 ones is regrouped as 2 tens and 4 ones.

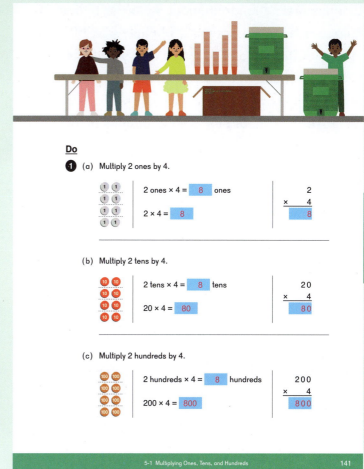

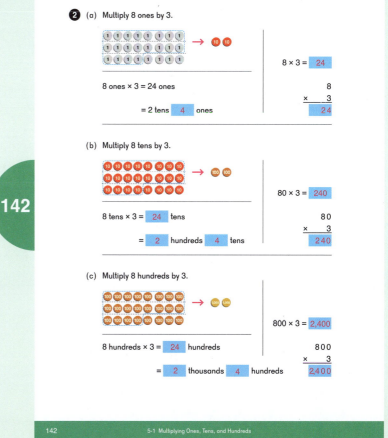

3 Discuss Mei's thoughts about the order of the numbers in the expression. "Multiply 4 by 500" means 500 groups of 4, which can be difficult to picture because of the large number of groups. It is easier to think of 4 groups of 500. We can do this because the product for both equations is the same.

Activity

▲ Spin and Multiply

Materials: Tens and Hundreds Spinner (BLM), Spin and Multiply Recording Sheet (BLM), pencil, paper clip

On each turn, a player spins the Tens and Hundreds Spinner (BLM) and records the numbers on the appropriate line of her Spin and Multiply Recording Sheet (BLM). She then multiplies each number and adds them together to get a score for that round. For example:

Player One spins 40 and 300 and writes the numbers on the × 2 line:

40 × 2 = 80   300 × 2 = 600   Score: 680

At the end of 4 rounds, players add their scores for each round to get a total. The player with the greatest total is the winner.

★ Rather than proceeding sequentially from × 2 to × 5, a player may place the numbers spun on any blank line to get the greatest total score.

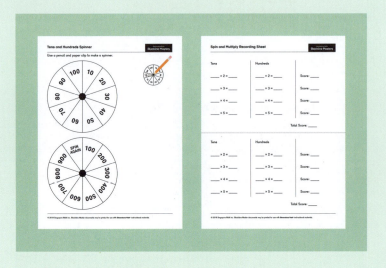

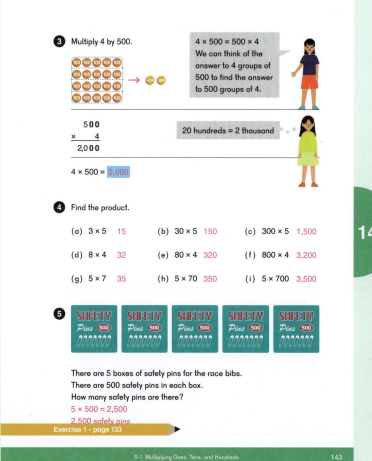

Exercise 1 • page 133

Lesson 2 Multiplication Without Regrouping

Objective

- Multiply a two-digit number by a one-digit number without regrouping.

Lesson Materials

- Place-value discs

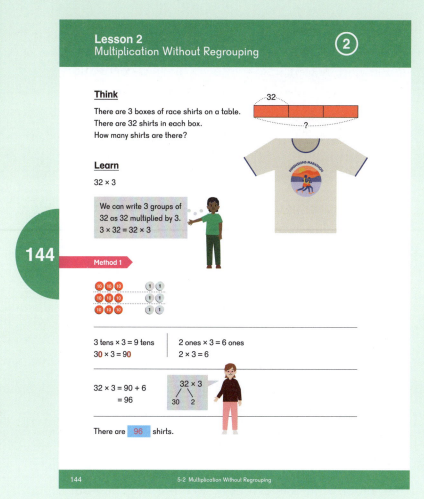

Think

Provide students with place-value discs and have them work the **Think** problem independently.

Have students write an expression to solve the problem and share how they found their answers.

Learn

Method 1 shows 32 split into tens and ones, which are each multiplied by 3, and the products added. This is a strategy many students can use mentally when the numbers are easy and there is not much regrouping.

Work through the **Think** problem with students as demonstrated in **Learn** for Method 2. Have students work along with place-value discs as the steps are modeled.

Have students discuss and compare the two methods in **Learn**.

They should note that with either Method 1 or Method 2, they end up with nine 10-discs and six 1-discs. They should see that both are ways of multiplying by place values.

Sofia is showing that instead of writing each product on a separate line, she can write the product from multiplying by tens immediately. This is a connection to the standard algorithm for students who may already be familiar with it. The standard algorithm will be developed further in the next lesson.

Ask students:

*   What is similar about the ways the problem has been solved in the textbook and by Sofia?
*   Which way is quickest and why? ("Sofia's way because you don't have to add again," or, "Sofia's way because we can do it mentally.")

Do

Students should use place-value discs if necessary. Most should be working with just the pages in the textbook.

❶—❷ The strategy of splitting numbers into the sum of their place values is extended to three-digit numbers.

Although this can be completed mentally, students can write down some of the partial products when needed:

$143 \times 2 = 200 + 80 + 6 = 286$

❸ Emphasize to students that Dion can write the sum immediately, instead of the product of the digit in each place.

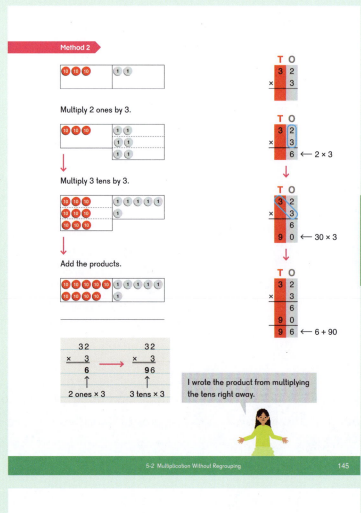

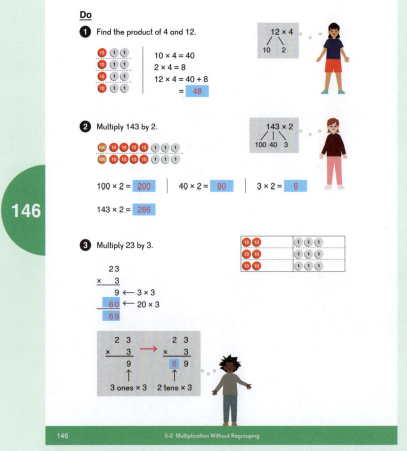

4 By thinking of 342 as 3 hundreds, 4 tens, and 2 ones, Alex can write the product in one row.

5 Students can use any of the methods they have learned. After they have worked **5** and **6**, have students share which methods they used for some of the problems and why they chose that method.

(f) Note that this problem has a 0 in the tens place.

Ask students:

- Do you need to write the partial product of 0 × 4?
- Can this problem be solved by thinking of the sum of 200 × 4 and 2 × 4?

Exercise 2 • page 135

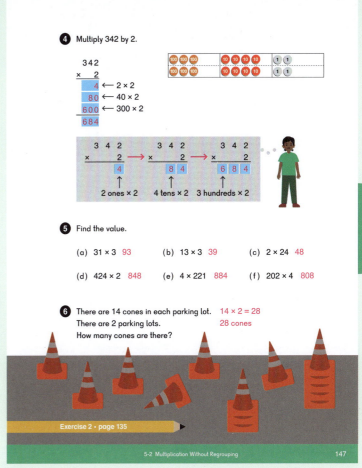

4 Multiply 342 by 2.

```
  3 4 2
×     2
      4  ← 2 × 2
    8 0  ← 40 × 2
  6 0 0  ← 300 × 2
  6 8 4
```

```
  3 4 2          3 4 2          3 4 2
×     2    →   ×     2    →   ×     2
      4            8 4          6 8 4
      ↑              ↑              ↑
2 ones × 2     4 tens × 2    3 hundreds × 2
```

5 Find the value.

(a) 31 × 3 93   (b) 13 × 3 39   (c) 2 × 24 48

(d) 424 × 2 848   (e) 4 × 221 884   (f) 202 × 4 808

6 There are 14 cones in each parking lot.   14 × 2 = 28
There are 2 parking lots.   28 cones
How many cones are there?

Exercise 2 • page 135

Lesson 3 Multiplication with Regrouping Tens

Objective

- Multiply a two-digit number by a one-digit number with regrouping tens.

Lesson Materials

- Place-value discs

Think

Provide students with place-value discs and have them work the **Think** problem independently.

Have students write equations and discuss how they found their answers.

Ask students:

- How is this problem different from the problems you solved in previous lessons? (We have to regroup tens.)
- How is it the same? (We can still multiply the digits in each place.)

Discuss student strategies for solving the problem. Ask them what they can do when they have more than 9 in the tens columns.

Ask students:

- Why do you think there is no bar model for this problem?
- Can we still draw a bar model?
- Do we need to draw a bar model with each problem?

(A model would show 42 units with 3 pies in each unit or 42 units of 3.)

Learn

Work through the **Think** problem with the students as demonstrated in **Learn**. Have students work along with place-value discs as the steps are modeled.

Discuss the regrouping from 12 tens to 1 hundred and 2 tens.

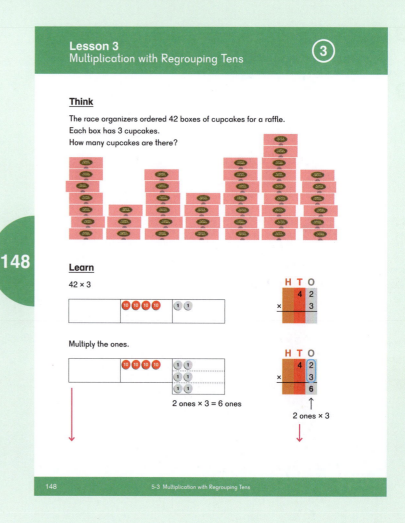

After the students have worked the problem with place-value discs, have them compare their methods from **Think** with the method shown in the textbook.

If needed, remind students that they know the answers to 40 × 3 and 2 × 3. They are learning how to the write the steps in a procedure that will help them with greater numbers.

Beginning with this lesson, the standard algorithm is shown first to explain how regrouping is recorded in the vertical equation. Students may continue to use partial products like Sofia or mental strategies like Mei if they can do so efficiently.

The standard algorithm (Emma's method) is efficient because we only multiply a single digit at a time.

The partial product method and place-value discs (Mei's way) are best for showing what is happening conceptually.

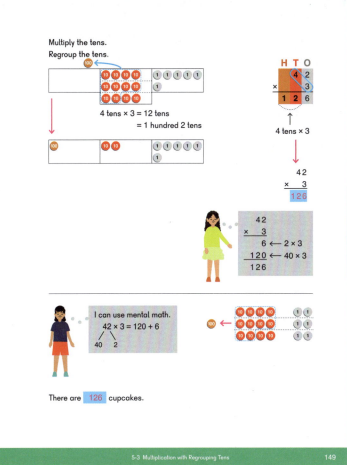

Do

Students should use place-value discs if necessary. Most should be working with just the pages in the textbook.

6 Students can use any of the methods they have learned. Have them share why they chose their methods after solving the problem.

Exercise 3 • page 137

150

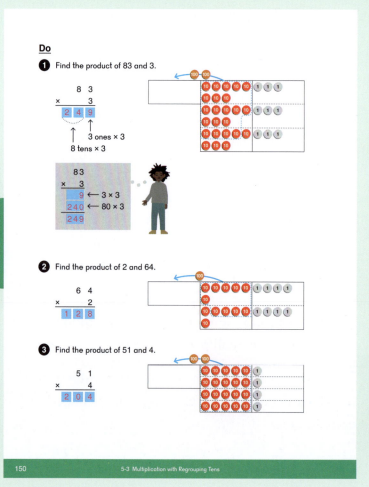

151

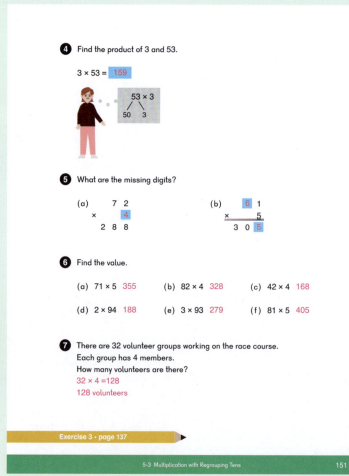

Lesson 4 Multiplication with Regrouping Ones

Objective

- Multiply a two-digit number by a one-digit number with regrouping ones.

Lesson Materials

- Place-value discs

Think

Provide students with place-value discs and have them work the **Think** problem independently.

Have students write an equation and discuss how they found their answers.

Ask students:

- How is this problem different from the ones you solved in the previous lesson? (We have to regroup ones.)
- How is it the same? (We can still multiply the digits in each place.)

Discuss student strategies for solving the problem. Ask them what they can do when they have more than 9 in the ones column.

Learn

Work through the **Think** problem with students as demonstrated in **Learn**. Have students work along with place-value discs as the steps are modeled.

Emphasize how to record the regrouped tens in the written algorithm, and the fact that this regrouped ten is not multiplied again when multiplying tens, but added in after multiplying tens.

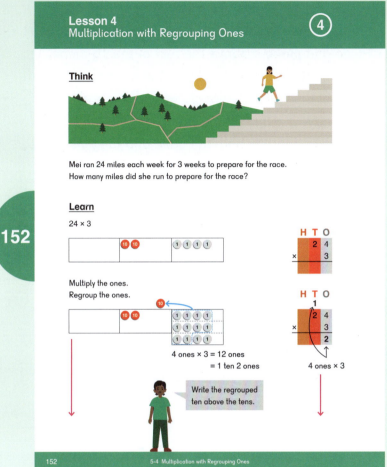

Discuss the regrouping from 12 ones to 1 ten and 2 ones. Ensure that students trade ten 1-discs for one 10-disc and place it above the rest of the 10-discs. Help students to understand that the regrouped tens do not get multiplied. Students who struggle with this will have a difficult time with problems where regrouping occurs in both the ones and the tens place.

Ask students:

- What is similar about the ways the problem has been solved by Dion, Emma, and Sofia?
- Whose way is quickest and why? ("Emma's method, because you don't have to add again," or, "Sofia's method, because we can do it mentally.")

After the students have worked the problem with place-value discs, have them compare their methods from **Think** with the method shown in the textbook.

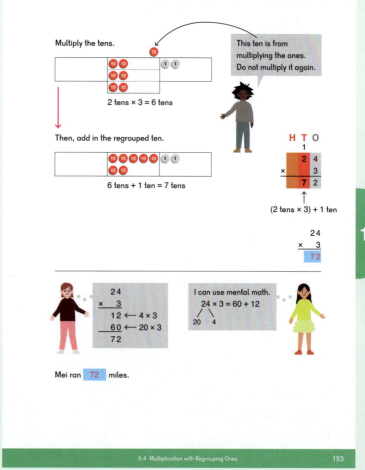

153

Do

❶ Struggling students may need to work these problems with place-value discs to see the regrouping step.

Mei reminds students not to multiply the regrouped tens.

❺ Students can use any of the methods they have learned. Have them share why they chose their methods after solving the problem.

Exercise 4 • page 139

154

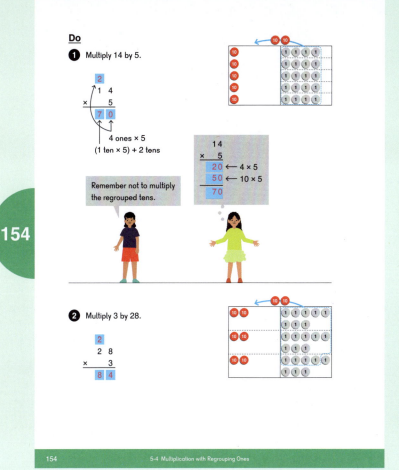

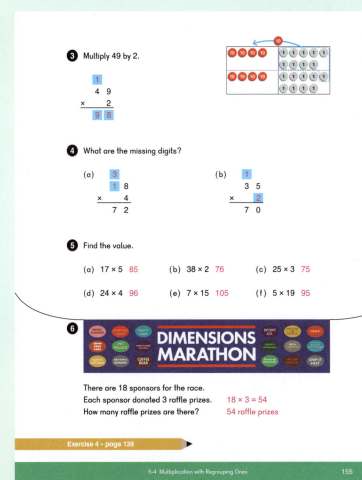

155

Lesson 5 Multiplication with Regrouping Ones and Tens

Objective

- Multiply a two-digit number by a one-digit number with regrouping tens and ones.

Lesson Materials

- Place-value discs

Think

Provide students with place-value discs and have them work the **Think** problem independently.

Ask students:

- How is this problem different from the ones you solved in the previous lesson? (We have to regroup ones and tens.)
- How is it the same? (We can still use place value to multiply.)

Discuss student strategies for solving the problem. Ask them what they can do when they have more than 9 in any column.

Learn

Work through the **Think** problem with students as demonstrated in **Learn**. Have students work along with place-value discs as the steps are modeled.

Discuss the regrouping from ones to tens and tens to hundreds.

Ensure that students trade ten 1-discs for one 10-disc and place it above the tens column. They should then trade ten 10-discs for one 100-disc and place it above the hundreds column. They should understand that the regrouped ten is only added after the original tens have been multiplied.

After the students have worked the problem with place-value discs, have them compare their methods from **Think** with the method shown in the textbook.

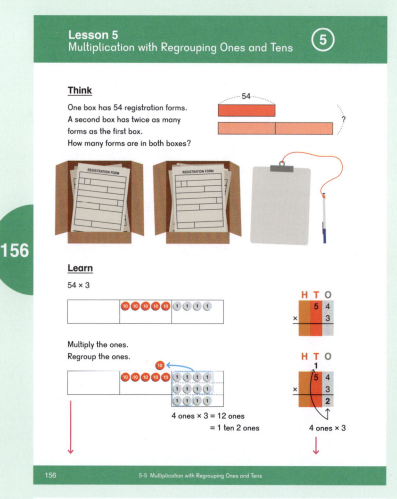

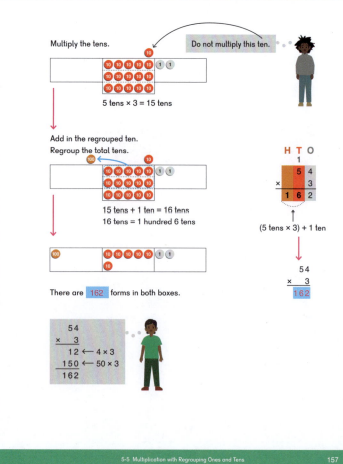

Do

Struggling students may need to work these problems with place-value discs to see the regrouping step.

② Sofia reminds students not to multiply the regrouped tens. Students can discuss Sofia's comment in relation to (3 tens × 5) + 2 tens in **①**.

⑤ Students can write the partial products and add if they still need to.

Provide struggling students with additional examples to work with place-value discs as needed.

Activity

★ How Many Can You Find? × 2

Materials: How Many Can You Find? × 2 (BLM), dry erase sleeves

Give each student a copy of How Many Can You Find? × 2 (BLM) in a dry erase sleeve. Using each digit 0 to 9 only once, challenge students to make a valid equation of a two-digit number multiplied by one-digit number, with a two-digit answer.

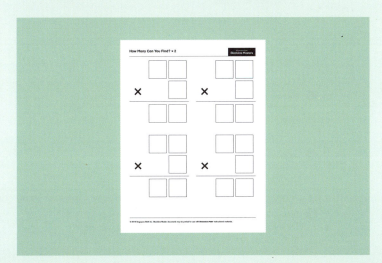

Exercise 5 • page 142

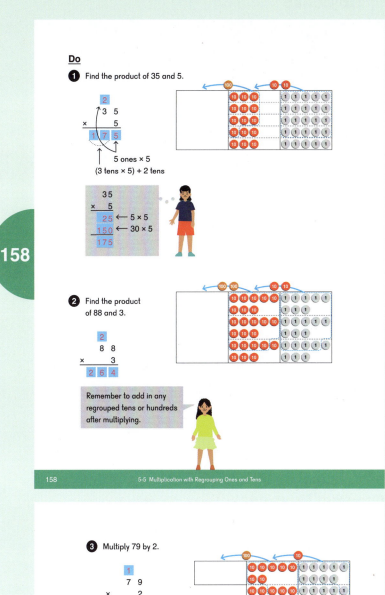

Do

① Find the product of 35 and 5.

$$\begin{array}{r} 3\,5 \\ \times \quad 5 \\ \hline 1\,7\,5 \end{array}$$

5 ones × 5
(3 tens × 5) + 2 tens

$$\begin{array}{r} 3\,5 \\ \times \quad 5 \\ \hline 2\,5 \\ 1\,5\,0 \\ \hline 1\,7\,5 \end{array}$$ ← 5 × 5
← 30 × 5

② Find the product of 88 and 3.

$$\begin{array}{r} 8\,8 \\ \times \quad 3 \\ \hline 2\,6\,4 \end{array}$$

Remember to add in any regrouped tens or hundreds after multiplying.

158   5-5 Multiplication with Regrouping Ones and Tens

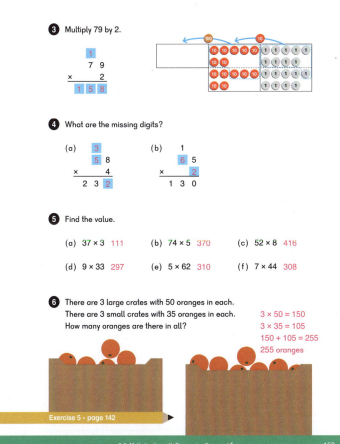

③ Multiply 79 by 2.

$$\begin{array}{r} 7\,9 \\ \times \quad 2 \\ \hline 1\,5\,8 \end{array}$$

④ What are the missing digits?

(a)
$$\begin{array}{r} 3 \\ 5\,8 \\ \times \quad 4 \\ \hline 2\,3\,2 \end{array}$$

(b)
$$\begin{array}{r} 1 \\ 6\,5 \\ \times \quad 2 \\ \hline 1\,3\,0 \end{array}$$

⑤ Find the value.

(a) 37 × 3 111   (b) 74 × 5 370   (c) 52 × 8 416

(d) 9 × 33 297   (e) 5 × 62 310   (f) 7 × 44 308

⑥ There are 3 large crates with 50 oranges in each.
There are 3 small crates with 35 oranges in each.
How many oranges are there in all?

3 × 50 = 150
3 × 35 = 105
150 + 105 = 255
255 oranges

Exercise 5 • page 142

5-5 Multiplication with Regrouping Ones and Tens   159

Objective

- Practice multiplying ones, tens, and hundreds by a one-digit number.

After students complete the **Practice** in the textbook, have them continue multiplying using activities from this chapter.

3—**10** Note that students' models may differ from the models in the textbook pages pictured in this guide. In addition, students may solve the problems in ways not included in this guide.

5—**7** and **10**, if represented by bar models, would show a large number of units at a number of dollars for each unit. As it is not practical for students to draw so many units, no bar model is given or expected from students.

Activity

▲ Greatest Product

Materials: Greatest Product — 2-digit Game Board (BLM), playing cards with face cards removed or Number Cards (BLM) 0 to 9

Players take turns drawing a card and placing it in one of the three places on their own Greatest Product — 2-digit Game Board (BLM). Players are trying to make an equation that has the greatest product. At any point, a player may discard one card.

After each player has created an equation, players multiply to find the answer, and the player with the greatest product is the winner.

Shuffle the cards and play again.

The goal could also be to create the least product.

Exercise 6 · page 145

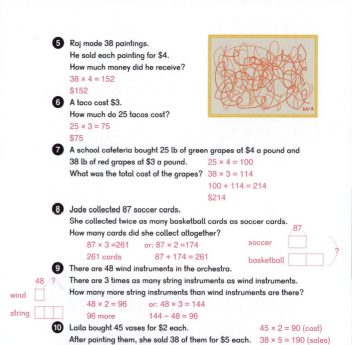

Lesson 7 Multiplying a 3-Digit Number with Regrouping Once

Objective

- Multiply a three-digit number with regrouping tens or ones.

Lesson Materials

- Place-value discs

Think

Provide students with place-value discs and have them work the **Think** problem independently.

Ask students:

- How is this problem different from the ones you solved in the previous lesson? (There are three digits.)
- How is it the same? (We are still multiplying.)

Discuss student strategies for solving the problem.

Learn

Work through the **Think** problem with students as demonstrated in **Learn**. Have students work along with place-value discs as the steps are modeled.

Discuss the regrouping from ones to tens.

Ensure that students trade ten 10-discs for one 100-disc and place it above the hundreds column. They should understand that the regrouped 100-disc does not get multiplied.

After the students have worked the problem with place-value discs, have them compare their methods from **Think** with the method shown in the textbook.

By this point, students' methods should be similar to textbook methods.

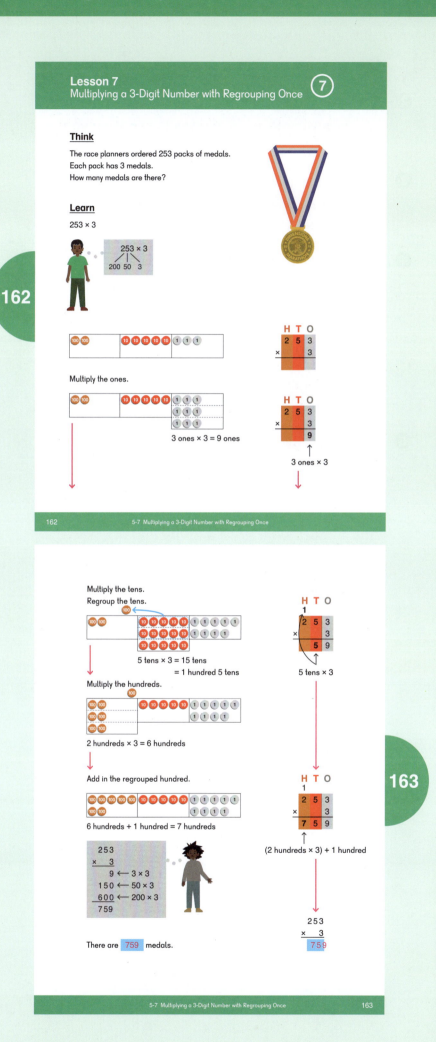

Do

② — **③** Struggling students may need to work these problems with place-value discs to see the regrouping step.

② Emma is thinking of the partial products method for this problem. Students may use this method until they are comfortable with the standard algorithm.

④ Students can use any of the methods they have learned. Have them share why they chose their methods after solving the problem. For example, problems similar to (d) could easily be solved mentally: 200 × 3 + 9 × 3.

Provide struggling students with additional examples for students to work with place-value discs as needed.

⑥ A bar model might be:

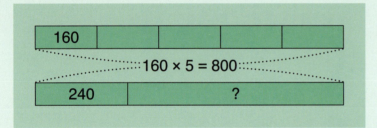

Exercise 7 • page 148

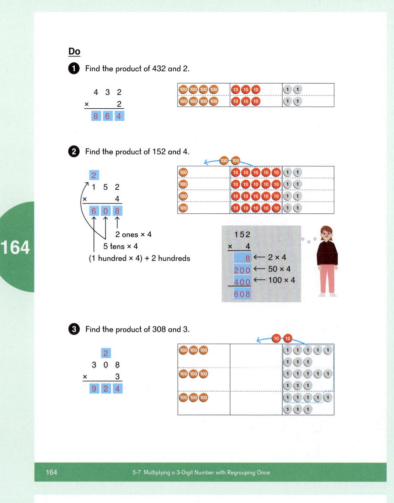

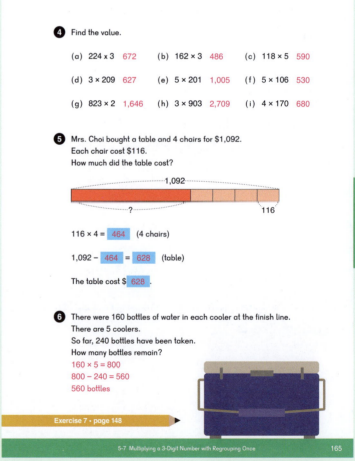

Lesson 8 Multiplication with Regrouping More Than Once

Objective

- Multiply a three-digit number with regrouping in more than one place.

Lesson Materials

- Place-value discs

Think

Provide students with place-value discs and have them work the **Think** problem independently.

Ask students:

- How is this problem different from the ones you solved in the previous lesson? (We have to regroup ones and tens and hundreds. The answer will be in the thousands.)
- How is it the same? (We can still multiply by the place value.)

Discuss student strategies for solving the problem. Ask them what they can do when they have more than 9 in any column.

Learn

Work through the **Think** problem with students as demonstrated in **Learn**. Have students work along with place-value discs as the steps are modeled.

Using partial products, this problem can be written as:

345 × 3 = 900 + 120 + 15

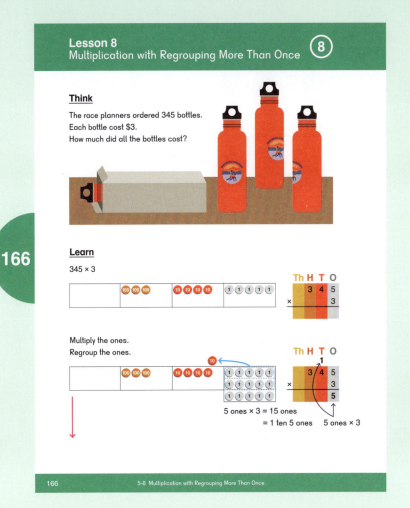

Discuss the regrouping and how it is recorded when multiplying ones, then tens, then hundreds.

Note that 3 hundreds × 3 + 1 more hundred does not necessarily need a mark above the thousands column. We can think of the number as 10 hundreds because there are no more digits to multiply.

Mei shows the partial products method for the problem.

After the students have worked the problem with place-value discs, have them compare their methods from **Think** with the method shown in the textbook.

Do

❶ — ❸ Students may need to work these problems with place-value discs to see the regrouping steps.

❶ Dion demonstrates the partial products method for this problem. Students may use this method until they are comfortable with the standard algorithm.

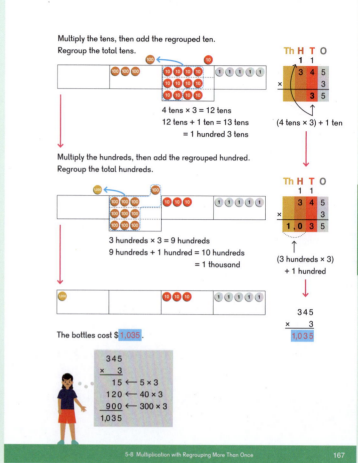

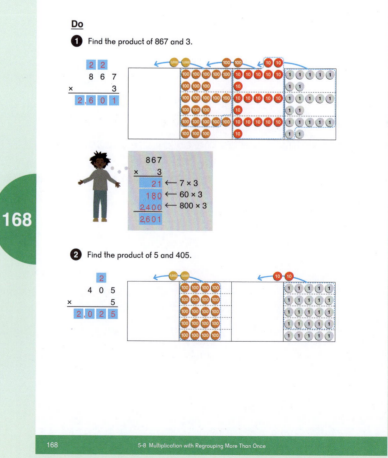

3 Challenge students to solve this with a different mental strategy: 999 is 1 less than 1,000.

2 thousands − 2 ones = 2,000 − 2 = 1,998

4 Students can use any of the methods they have learned. Have them share why they chose their methods after solving the problem.

Activity

★ **How Many Can You Find? × 3**

Materials: How Many Can You Find? × 3 (BLM), dry erase sleeves

Using How Many Can You Find? × 3 (BLM) in dry erase sleeves, extend the activity from Lesson 5 on page 188 of this Teacher's Guide. Challenge students to create a valid equation in which a three-digit number is multiplied by a one-digit number resulting in a three-digit product.

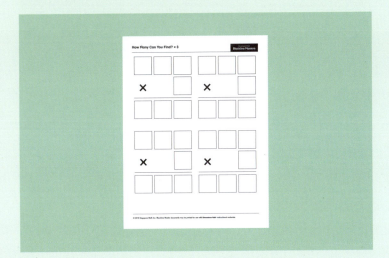

Exercise 8 • page 151

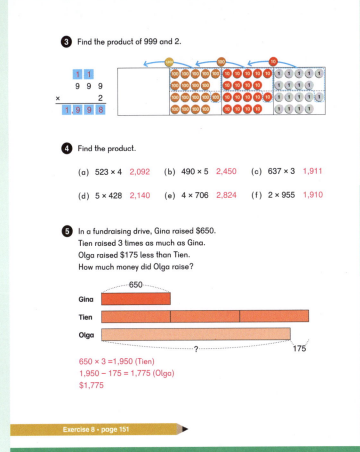

Lesson 9 Practice B

Objective

- Practice multiplying three-digit numbers.

Activity

▲ Greatest Product

Materials: Greatest Product — 3-digit Game Board (BLM), playing cards with face cards removed or Number Cards (BLM) 0 to 9

Players take turns drawing a card and placing it in one of the three places on their own Greatest Product — 3-digit Game Board (BLM). Players are trying to make an equation that has the greatest product. At any point, a player may discard one card into the trash.

After each player has created an equation, players multiply to find the answer, and the player with the greatest product is the winner.

Shuffle the cards and play again.

The goal could also be to create the least product.

Exercise 9 • page 154

Brain Works

★ Make it True

Materials: 1 set of Number Cards (BLM) 0 to 9, Multiplication Symbol Cards (BLM)

Using each card only once, make as many true equations as you can.

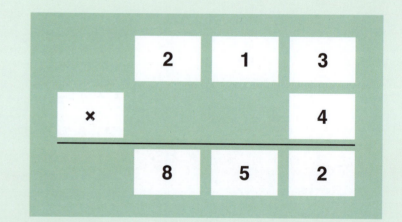

Chapter 5 Multiplication

Exercise 1

Basics

1 Fill in the missing numbers or digits.

(a) 7 ones × 4 = [28] ones

= [28]

```
      7
  ×   4
  [2][8]
```

(b) 7 tens × 4 = [28] tens

= [280]

```
      7 0
  ×     4
  [2][8]0
```

(c) 7 hundreds × 4 = [28] hundreds

= [2,800]

```
      7 0 0
  ×       4
  [2,][8][0]0
```

2 Multiply.

(a)

```
      2
  ×   4
  [8]
```

```
    2 0
  ×   4
  [8][0]
```

```
    2 0 0
  ×     4
  [8][0][0]
```

(b)
```
      5
  ×   4
  [2][0]
```
```
    5 0
  ×   4
  [2][0][0]
```

```
    5 0 0
  ×     4
  [2,][0][0][0]
```

Practice

3 (a) Multiply 3 by 600.

```
      6 0 0
  ×       3
  [1,][8][0][0]
```

(b) Find the product of 80 and 5.

```
      8 0
  ×     5
  [4][0][0]
```

4 (a) 20 × 5 = [100] (b) 700 × 2 = [1,400]

(c) 4 × 90 = [360] (d) 4 × 600 = [2,400]

(e) 3 × 800 = [2,400] (f) 800 × 5 = [4,000]

5 There are 4 boxes of 800 nails.
How many nails are there in all?
800 × 4 = 3,200
There are 3,200 nails in all.

Challenge

6 There are 300 packages each with a set of 5 screwdrivers, and 500 packages each with a set of 4 screwdrivers. How many screwdrivers are there in all?
300 × 5 = 1,500
500 × 4 = 2,000
1,500 + 2,000 = 3,500
There are 3,500 screwdrivers in all.

Exercise 2

Basics

1 Fill in the missing numbers or digits.

2 × 3 = [6]

30 × 3 = [90]

100 × 3 = [300]

132 × 3 = [396]

```
      1 3 2
  ×       3
        [6]   ← 2 × 3
      [9][0]  ← 30 × 3
    [3][0][0] ← 100 × 3
    [3][9][6]
```

```
      1 3 2
  ×       3
    [3][9][6]
         ↑  ↑  ↑
              2 × 3
           30 × 3
        100 × 3
```

2 Multiply.

```
    2 1
  ×   4
  [8][4]
```
```
    1 2
  ×   4
  [4][8]
```
```
    1 1 2
  ×     3
  [3][3][6]
```

3 Find the value of 2 × 321.

```
    3 2 1
  ×     2
  [6][4][2]
```

Practice

4 Multiply.

33 × 3
```
      3 3
  ×     3
  [9][9]
```

141 × 2
```
    1 4 1
  ×     2
  [2][8][2]
```

2 × 432
```
    4 3 2
  ×     2
  [8][6][4]
```

5 A craft store sells wooden beads in bags of 230.
How many beads are in two bags?
2 × 230 = 460
There are 460 beads in two bags.

Challenge

6 Aki made 12 pairs of earrings.
She used 2 beads on each earring.
How many beads did she use in all?
2 × 2 = 4   or   12 × 2 = 24
12 × 4 = 48   24 × 2 = 48

She used 48 beads in all.

Teacher's Guide 3A Chapter 5

Basics

1 Fill in the missing numbers or digits.

$2 \times 3 = \boxed{6}$

$80 \times 3 = \boxed{240}$

$82 \times 3 = \boxed{246}$

```
      8 2
  ×     3
  ───────
        6  ← 2 × 3
    2 4 0  ← 80 × 3
  ───────
    2 4 6
```

```
      8 2
  ×     3
  ───────
    2 4 6
       ↑
       2 × 3
    80 × 3
```

2 (a) $74 \times 2 = 140 + 8 = \boxed{148}$

    70 4

(b) $5 \times 61 = \boxed{300} + 5 = \boxed{305}$

    $\boxed{60}$ 1

(c) $93 \times 3 = \boxed{270} + 9 = \boxed{279}$

    $\boxed{90}$ 3

Practice

3 Multiply.

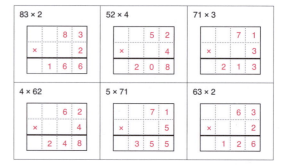

83 × 2	52 × 4	71 × 3
8 3 / × 2 / **1 6 6**	5 2 / × 4 / **2 0 8**	7 1 / × 3 / **2 1 3**

4 × 62	5 × 71	63 × 2
6 2 / × 4 / **2 4 8**	7 1 / × 5 / **3 5 5**	6 3 / × 2 / **1 2 6**

4 A deck of cards has 52 cards.
How many cards are in 4 decks of cards?

$4 \times 52 = 208$

There are 208 cards in 4 decks of cards.

Challenge

5 Write the missing digits in each equation to make them true.
The missing numbers can be used more than once.

(a) Missing: 3 and 5

```
      5 2
  ×     3
  ───────
    1 5 6
```

(b) Missing: 2 and 4

```
      8 2
  ×     4
  ───────
    3 2 8
```

Exercise 4

Basics

Fill in the missing numbers or digits.

1 $8 \times 3 = \boxed{24}$

$20 \times 3 = \boxed{60}$

$28 \times 3 = \boxed{84}$

```
    2 8
  ×   3
  ┌─────┐
  │2 4│ ← 8 × 3
  │6 0│ ← 20 × 3
  └─────┘
  │8 4│
```

```
    │2│
    2 8
  ×   3
  ┌───┐
  │8│4│
    ↑  ↑
  8 × 3
  20 × 3 + 20
```

2 $8 \times 2 = \boxed{16}$

$38 \times 2 = 30 \times 2 + \boxed{16}$

$= \boxed{76}$

```
    3 8
  ×   2
  ┌─────┐
  │7 6│
```

3 $9 \times 4 = \boxed{36}$

$19 \times 4 = 10 \times 4 + \boxed{36}$

$= \boxed{76}$

```
    1 9
  ×   4
  ┌─────┐
  │7 6│
```

Practice

4 Multiply.

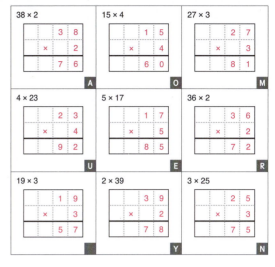

38 × 2

		3	8
	×		2
		7	6

A

15 × 4

		1	5
	×		4
		6	0

O

27 × 3

		2	7
	×		3
		8	1

M

4 × 23

		2	3
	×		4
		9	2

U

5 × 17

		1	7
	×		5
		8	5

E

36 × 2

		3	6
	×		2
		7	2

R

19 × 3

		1	9
	×		3
		5	7

2 × 39

		3	9
	×		2
		7	8

Y

3 × 25

		2	5
	×		3
		7	5

N

Riddle: What belongs to you but is used more by others?
Write the letters (or blank for space) in the boxes below to find out.

Y	O	U	R		N	A	M	E
78	60	92	72	57	75	76	81	85

5 Each student in the class is to be given 4 sheets of paper.
If there are 24 students in the class, how many sheets of paper
are needed?

$24 \times 4 = 96$

96 sheets of paper are needed.

6 A gift basket has 12 kiwis and 12 tangerines.
How many fruits do 2 such gift baskets have?

$12 + 12 = 24$

$24 \times 2 = 48$

2 such gift baskets have 48 fruits.

Challenge

7 $19 \times 4 = 20 \times 4 - \boxed{4}$

8 Write the missing digits in each equation to make them true.

(a) Missing: 1, 5, and 6

```
      1 6
  ×     5
  ───────
    8 0
```

(b) Missing: 1, 2, and 3

```
      2 7
  ×     3
  ───────
    8 1
```

Teacher's Guide 3A Chapter 5

Exercise 5

Basics

1 Fill in the missing numbers or digits.

$8 \times 3 = \boxed{24}$

$60 \times 3 = \boxed{180}$

$68 \times 3 = \boxed{204}$

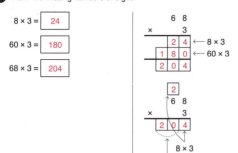

$$\begin{array}{r} 6\ 8 \\ \times \quad 3 \\ \hline \boxed{2\ 4} \leftarrow 8 \times 3 \\ \boxed{1\ 8\ 0} \leftarrow 60 \times 3 \\ \hline \boxed{2\ 0\ 4} \end{array}$$

$$\begin{array}{r} \boxed{2} \\ 6\ 8 \\ \times \quad 3 \\ \hline 2\ \boxed{0}\ \boxed{4} \end{array}$$
8 × 3
60 × 3 + 20

2 $8 \times 5 = \boxed{40}$

$38 \times 5 = 30 \times 5 + \boxed{40}$

$= \boxed{190}$

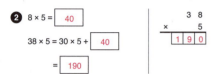

$$\begin{array}{r} 3\ 8 \\ \times \quad 5 \\ \hline 1\ \boxed{9}\ \boxed{0} \end{array}$$

3 $9 \times 4 = \boxed{36}$

$79 \times 4 = 70 \times 4 + \boxed{36}$

$= \boxed{316}$

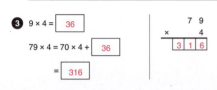

$$\begin{array}{r} 7\ 9 \\ \times \quad 4 \\ \hline \boxed{3}\ 1\ \boxed{6} \end{array}$$

Practice

4 Multiply.

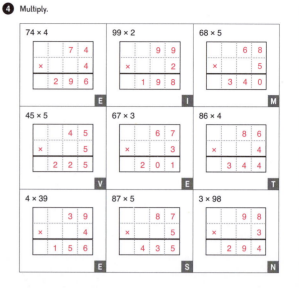

74 × 4	99 × 2	68 × 5
296 **E**	198 **I**	340 **M**

45 × 5	67 × 3	86 × 4
225 **V**	201 **E**	344 **T**

4 × 39	87 × 5	3 × 98
156 **E**	435 **S**	294 **N**

What is the greatest number of times you can fold ordinary printer paper in half by hand?
Write the letters in the boxes below to find out.

S	E	V	E	N		T	I	M	E	S
435	201	225	156	294	286	344	198	340	296	435

5 Isabella made 5 bows.
She used 86 cm of ribbon for each bow.
How many centimeters of ribbon did she use altogether?
$5 \times 86 = 430$
She used 430 cm of ribbon.

6 A store received 4 crates of grape juice one week and
3 crates of grape juice the next week.
Each crate had 75 bottles of grape juice.
How many bottles of grape juice did the store receive those two weeks?
$4 + 3 = 7$
$7 \times 75 = 525$
The store received 525 bottles of grape juice those two weeks.

Challenge

7 $99 \times 4 = 100 \times 4 - \boxed{4}$

8 Write the missing digits in each equation to make them true.

(a) Missing: 2 and 4

$$\begin{array}{r} 8\ \boxed{2} \\ \times \quad 6 \\ \hline \boxed{4}\ 9\ 2 \end{array}$$

(b) Missing: 6 and 7

$$\begin{array}{r} \boxed{6}\ 3 \\ \times \quad 9 \\ \hline 5\ \boxed{6}\ \boxed{7} \end{array}$$

Exercise 6

<u>Check</u>

1 (a) 800 × 5 = [4,000]

(b) 3 × 900 = [2,700]

(c) 4 × 40 = [160]

(d) 600 × 4 = [2,400]

(e) 3 × 200 = [600]

(f) 500 × 5 = [2,500]

(g) 32 × 3 = [96]

(h) 122 × 4 = [488]

2 Multiply.

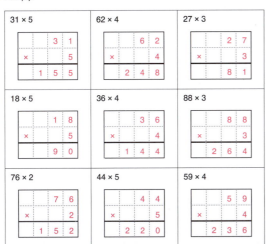

31 × 5	62 × 4	27 × 3
3 1 × 5 = 1 5 5	6 2 × 4 = 2 4 8	2 7 × 3 = 8 1
18 × 5	36 × 4	88 × 3
1 8 × 5 = 9 0	3 6 × 4 = 1 4 4	8 8 × 3 = 2 6 4
76 × 2	44 × 5	59 × 4
7 6 × 2 = 1 5 2	4 4 × 5 = 2 2 0	5 9 × 4 = 2 3 6

3 A bakery uses 5 cups of flour to make one chocolate cake,
4 cups of flour to make one vanilla cake,
and 3 cups of flour to make one pudding cake.

(a) How many cups of flour does the bakery need to make 23 chocolate
cakes and 32 vanilla cakes?
23 × 5 = 115
32 × 4 = 128
115 + 128 = 243
The bakery needs 243 cups of flour.

(b) Which takes more flour to make, 16 chocolate cakes or 25 pudding cakes?
How much more?

16 × 5 = 80
25 × 3 = 75
It takes 5 more cups of flour to make 16 chocolate cakes
than to make 25 pudding cakes.

(c) A baker combined the recipes for vanilla cake and pudding cake to make
a large vanilla-pudding cake.
How many cups of flour are needed to make 22 vanilla-pudding cakes?
3 + 4 = 7
22 × 7 = 154
154 cups of flour are need to make 22 vanilla-pudding cakes.

4 Gavin collected 43 game cards.
Carlos collected three times as many game cards as Gavin.
How many cards did they collect altogether?

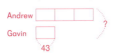

Andrew [ ][ ][ ] ?
Gavin [ ]
43

43 × 4 = 172
They collected 172 cards altogether.

<u>Challenge</u>

5 Madison has twice as many game cards as Imani.
Imani has 27 more game cards than Kiara.
Kiara has 55 game cards.
How many game cards do they have altogether?

Madison | 55 | 27 | 55 | 27 |
Imani | 55 | 27 |
Kiara | 55 |

55 × 4 = 220
27 × 3 = 81
220 + 81 = 301

The have 301 game cards altogether.

6 Arrange the digits 3, 4, and 5 to form one number with the greatest product
and one number with the least product.

4 3
× 5
2 1 5

4 5
× 3
1 3 5

Exercise 7

Basics

1 Fill in the missing numbers or digits.

$2 \times 3 =$ 6

$40 \times 3 =$ 120

$100 \times 3 =$ 300

$142 \times 3 =$ 426

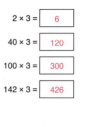

```
    1 4 2
  ×     3
        6   ← 2 × 3
    1 2 0   ← 40 × 3
    3 0 0   ← 100 × 3
    4 2 6
```

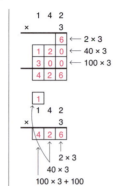

```
      1
    1 4 2
  ×     3
    4 2 6
          2 × 3
        40 × 3
     100 × 3 + 100
```

2 Multiply.

118 × 4
```
      1 1 8
  ×       4
      4 7 2
```

182 × 4
```
      1 8 2
  ×       4
      7 2 8
```

812 × 4
```
      8 1 2
  ×       4
    3, 2 4 8
```

Practice

3 Multiply.

317 × 2
```
      3 1 7
  ×       2
      6 3 4
```
S

5 × 171
```
      1 7 1
  ×       5
      8 5 5
```
E

123 × 4
```
      1 2 3
  ×       4
      4 9 2
```
I

4 × 512
```
      5 1 2
  ×       4
    2, 0 4 8
```
E

513 × 3
```
      5 1 3
  ×       3
    1, 5 3 9
```
T

924 × 2
```
      9 2 4
  ×       2
    1, 8 4 8
```
M

307 × 3
```
      3 0 7
  ×       3
      9 2 1
```
W

2 × 263
```
      2 6 3
  ×       2
      5 2 6
```
V

171 × 4
```
      1 7 1
  ×       4
      6 8 4
```
L

A student set a record by folding a long piece of toilet paper in half...
Write the letters in the boxes below to complete the sentence.

T	W	E	L	V	E		T	I	M	E	S
1,539	921	855	684	526	2,048	426	1,539	492	1,848	2,048	634

4 A drier costs $412.
A washing machine costs twice as much.
How much do the washing machine and dryer cost altogether?
$412 × 3 = $1,236
They cost $1,236 altogether.

5 Karen bought a dining room table and 4 chairs.
The chairs each cost $115.
The table cost $532.
How much did she spend?
$115 × 4 = $460
$460 + $532 = $992
She spent $992.

Challenge

6 Write the missing digits in each equation to make them true.

(a) Missing: 2 and 6
```
      6 2 2
  ×       3
    1, 8 6 6
```

(b) Missing: 0, 1, and 5

```
      1 3 0
  ×       5
      6 5 0
```

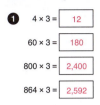

Exercise 8

Basics

Fill in the missing numbers or digits.

1 4 × 3 = **12**

60 × 3 = **180**

800 × 3 = **2,400**

864 × 3 = **2,592**

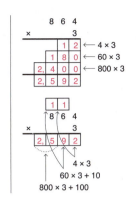

```
        8 6 4
    ×       3
        1 2   ← 4 × 3
      1 8 0   ← 60 × 3
    2 4 0 0   ← 800 × 3
    2 5 9 2
```

```
      1 1
      8 6 4
    ×     3
    2 5 9 2
              4 × 3
              60 × 3 + 10
          800 × 3 + 100
```

2 5 × 5 = **25**

40 × 5 = **200**

900 × 5 = **4,500**

945 × 5 = **4,725**

```
      2 2
      9 4 5
    ×     5
    4 7 2 5
              5 × 5
              40 × 5 + 20
          900 × 5 + 200
```

Practice

3 Multiply.

728 × 4	684 × 2	3 × 359
7 2 8 × 4 = 2,912	6 8 4 × 2 = 1,368	3 5 9 × 3 = 1,077
T	**E**	**I**

789 × 5	4 × 369	408 × 5
7 8 9 × 5 = 3,945	3 6 9 × 4 = 1,476	4 0 8 × 5 = 2,040
R	**H**	**S**

3 × 665	427 × 3	564 × 4
6 6 5 × 3 = 1,995	4 2 7 × 3 = 1,281	5 6 4 × 4 = 2,256
V	**N**	**U**

If a piece of paper could be folded 103 times, it would be thicker than...
Write the letters in the boxes below to complete the sentence.

T	H	E		U	N	I	V	E	R	S	E
2,912	1,476	1,368		1,446	2,256	1,281	1,077	1,995	1,368	3,945	2,040	1,368

4 A bicycle costs $519.
A motorcycle costs 5 times as much as the bicycle.
How much more does the motorcycle cost than the bicycle?
$519 × 4 = $2,076
The motorcycle costs $2,076 more than the bicycle.

5 Malik had $3,500 to spend on some office furniture.
He bought 3 desks. Each desk costs $999.
How much money does he have left?
$999 × 3 = $2,997
$3,500 − $2,997 = $503
He has $503 left.

Challenge

6 Write the missing digits in each equation to make them true.

(a) Missing: 1, 3, and 7

```
      3 6 7
    ×     3
    1 1 0 1
```

(b) Missing: 2 and 5

```
      5 5 5
    ×     4
    2 2 2 0
```

Teacher's Guide 3A Chapter 5

Exercise 9

Check

1 Multiply.

143 × 3	2 × 739	690 × 4
1 4 3 × 3 4 2 9	7 3 9 × 2 1, 4 7 8	6 9 0 × 4 2, 7 6 0

5 × 174	605 × 4	824 × 3
1 7 4 × 5 8 7 0	6 0 5 × 4 2, 4 2 0	8 2 4 × 3 2, 4 7 2

999 × 2	555 × 5	4 × 666
9 9 9 × 2 1, 9 9 8	5 5 5 × 5 2, 7 7 5	6 6 6 × 4 2, 6 6 4

2 Find the missing digits.

(a)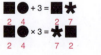
$$\begin{array}{r} 1\ 7\ 2 \\ \times \quad \boxed{5} \\ \hline \boxed{8}\ 6\ 0 \end{array}$$

(b)
$$\begin{array}{r} \boxed{4}\ \boxed{5}\ 9 \\ \times \quad\quad 3 \\ \hline 1,\ 3\ 7\ 7 \end{array}$$

3 A hotel has 452 rooms.

(a) Each room has 3 chairs.
How many chairs are in all the rooms?
452 × 3 = 1,356
There are 1,356 chairs in all the rooms.

(b) There are 2 beds in 355 rooms and 1 bed in the rest of the rooms.
How many beds are there altogether?
452 − 355 = 97
355 × 2 = 710
710 + 97 = 807
There are 807 beds altogether.

(c) A room costs $159 a night Sunday through Thursday and $178 a night Friday and Saturday.
What does it cost to stay at the hotel for a whole week?
$159 × 5 = $795
$178 × 2 = $356
$795 + $356 = $1,151
It costs $1,151 to stay at the hotel for a whole week.

Challenge

4 The Chens stayed at the hotel for 3 nights.
They rented 2 rooms with two beds and 1 room with a single bed.
The room with two beds costs $159 a night and the room with a single bed costs $172 a night.
How much did they spend?
$159 × 2 × 3 = $159 × 6 = $954
$172 × 3 = $516
$954 + $516 = $1,470
They spent $1,470.

5 Each symbol stands for a different digit.
Find the digits.

$$\blacksquare \bullet + 3 = \blacksquare \text{✱}$$
$$\underset{2\ 4}{\blacksquare \bullet} \quad \underset{2\ 7}{\blacksquare \text{✱}}$$
$$\blacksquare \bullet \times 3 = \text{✱} \blacksquare$$
$$\underset{2\ 4}{\blacksquare \bullet} \quad \underset{7\ 2}{\text{✱} \blacksquare}$$

$\blacksquare = \boxed{2}$

$\bullet = \boxed{4}$

$\text{✱} = \boxed{7}$

• ■ must be 1, 2, or 3, since multiplying ■ tens by 3 is a 2-digit number.
• ● must be less than 7, since adding 3 does not change the tens digit.
• ● × 3 has to have 1, 2, or 3 (■) in the ones place.
So ● could be 1 or 4 and ■ can't be 1.
Try 21, 24, 31, 34. Only 24 works.

Suggested number of class periods: 9–10

	Lesson	Page	Resources		Objectives
	Chapter Opener	p. 209	TB:	p. 171	Investigate division.
1	Dividing Tens and Hundreds	p. 210	TB: WB:	p. 172 p. 157	Divide tens or hundreds by 1 one-digit number when there is no remainder. Become familiar with the long division symbol and relate the positions of the dividend, divisor, and quotient to the equation form.
2	Dividing a 2-Digit Number by 2 — Part 1	p. 212	TB: WB:	p. 174 p. 159	Divide a number of up to two digits by 2 when there is a remainder.
3	Dividing a 2-Digit Number by 2 — Part 2	p. 215	TB: WB:	p. 178 p. 161	Divide a two-digit number by 2 when there is a remainder for tens.
4	Dividing a 2-Digit Number by 3, 4, and 5	p. 218	TB: WB:	p. 182 p. 164	Divide a two-digit number by 3, 4, and 5.
5	Practice A	p. 221	TB: WB:	p. 186 p. 167	Practice division.
6	Dividing a 3-Digit Number by 2	p. 223	TB: WB:	p. 188 p. 170	Divide three-digit numbers by 2 when there is regrouping in the tens and/or ones.
7	Dividing a 3-Digit Number by 3, 4, and 5	p. 226	TB: WB:	p. 192 p. 174	Divide three-digit numbers by 3, 4, and 5 when there is regrouping in the hundreds, tens, and ones.
8	Dividing a 3-Digit Number, Quotient is 2 Digits	p. 229	TB: WB:	p. 196 p. 177	Divide a three-digit number when the digit in the hundreds place cannot be divided by the divisor.
9	Practice B	p. 232	TB: WB:	p. 199 p. 180	Practice division.
	Workbook Solutions	p. 234			

This chapter introduces the standard long division algorithm. This algorithm begins by dividing the highest place digit in the dividend by the divisor. Any remainder in that place is regrouped one place to the right. This is repeated until the entire number has been divided. At this level, the final remainder is left as a whole number that is "left over."

The division algorithm is a complex process. It requires an understanding of all four operations, as well as place value and regrouping. In addition, it looks different from the standard addition, subtraction, and multiplication algorithms.

Students will use place-value discs to enrich their conceptual understanding of the algorithm. Students should use place-value discs when first solving the **Think** problems as they gain procedural fluency.

Lessons begin with simple division problems to introduce the notation. Then they move to dividing two-digit numbers by 2 to reinforce the concept of a remainder, and then by 3, 4, and 5.

All lessons use manipulatives which students should be encouraged to use to understand the division algorithm. Students need to understand the steps in the algorithm conceptually before moving on to **Dimensions Math® 3B Chapter 8: Multiplying and Dividing with 6, 7, 8, and 9**, where the algorithm will be practiced without place-value discs. Lessons in this chapter may take more than one class period. Time spent here will be well served in future chapters.

Lessons will show the two or three-digit number split into the values of the individual digits, or a part that can be divided evenly by the divisor and the remainder to emphasize that the digit in each place is treated as a unit. Students can also use this idea to solve an easy division problem mentally.

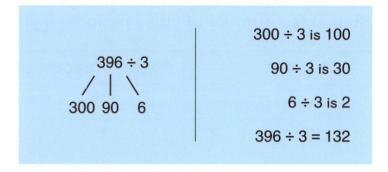

Standard Algorithm

While division can be understood as both sharing and grouping, this algorithm can best be understood as sharing division. The general procedure for the division algorithm with place-value discs is given here.

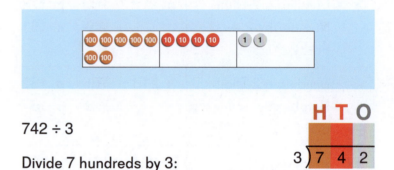

$742 \div 3$

Divide 7 hundreds by 3:

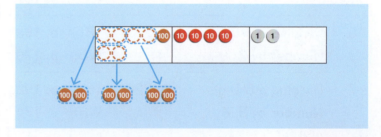

- 7 hundreds divided into 3 equal groups is 2 hundreds in each group with 1 hundred remaining. Write the digit 2 in the hundreds column above the line.
- Of the 7 hundreds, 6 were divided equally into groups. 7 hundreds − 6 hundreds = 1 hundred, which remains to be divided.
- Regroup the remaining 1 hundred as 10 tens. There are now 14 tens.

Divide 14 tens by 3:

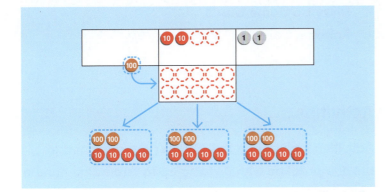

- 14 tens divided into 3 equal groups is 4 tens in each group with 2 tens remaining. Write the digit 4 in the tens column above the line.
- Of the 14 tens, 12 were divided equally into groups. 14 tens − 12 tens = 2 tens, which remain to be divided.
- Regroup the remaining 2 tens as 20 ones. There are now 22 ones.

$$\begin{array}{r} 2\ 4 \\ 3\overline{)7\ 4\ 2} \\ 6 \\ \overline{1\ 4} \\ 1\ 2 \\ \overline{2} \end{array}$$

Divide 22 ones by 3:

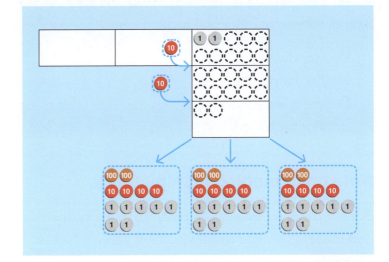

- 22 ones divided into 3 equal groups is 7 ones in each group with 1 one remaining. Write the digit 7 in the ones column above the line.
- Of the 22 ones, 21 were divided equally into groups. 22 ones − 21 ones = 1, so there is a remainder of 1.

$$\begin{array}{r} 2\ 4\ 7 \\ 3\overline{)7\ 4\ 2} \\ 6 \\ \overline{1\ 4} \\ 1\ 2 \\ \overline{2\ 2} \\ 2\ 1 \\ \overline{1} \end{array}$$

In 742 ÷ 3 = 247 × 3 + 1, the quotient is 247 and the remainder is 1. This can be written as 247 R 1.

Bar models are given when appropriate, however, they can be challenging to draw with a remainder.

When teaching the division algorithm, emphasize place value. Concentrate on the reasons for the steps in the long division algorithm and avoid mnemonics like:

- Dad for divide
- Mother for multiply
- Sister for subtract
- Brother for bring down

Note on managing manipulatives

Base ten manipulatives are pictured in the textbook and students should use them when solving division problems initially.

Most students will understand long division with base ten blocks and place-value discs.

Due to the amount of manipulatives used in this chapter, consider structuring the lessons in the following way:

Students should use the manipulatives first to understand how multi-digit division works. Teachers can then demonstrate the algorithm while students work with manipulatives. Next, students work the algorithm and use discs at the same time. Finally, the students no longer require the disc and will simply work with the algorithm.

The goal is to have the students work the problems without place-value discs.

Materials

- Bags
- Base ten blocks (tens and ones)
- Counters
- Die with modified sides: 0 in place of 6
- Die with modified sides: "roll again" in place of 6
- Division fact cards (students will make these in the Chapter Opener)
- Index cards
- Place-value discs
- Recording sheet

Blackline Masters

- Divvy Up Game Cards
- Kaboom Cards
- Number Cards

Activities

Activities included in this chapter are designed to provide practice with dividing up to three-digit numbers. They can be used after students complete the **Do** questions, or any time additional practice is needed.

Objective

- Investigate division.

Lesson Materials

- Index cards, 50 per student

Provide each student with 50 index cards to create their own flash cards for division facts for 2, 3, 4, 5, and 10. These cards will be used in the activity Division Kaboom.

Have students lay the cards out and look for patterns in the dividends, divisors, and quotients.

Review multiplication facts for 2, 3, 4, 5, and 10 as a warm-up and play any games from **Chapter 4: Multiplication and Division**.

Activity

▲ Division Kaboom

Materials: Division fact cards made by students, Kaboom Cards (BLM)

Shuffle fact cards and 3 Kaboom Cards (BLM) together, and place them facedown in a pile. Players take turns drawing a card and saying the answer to the division fact.

If a student answers correctly, she keeps the card. If she answers incorrectly, the card is returned to the pile. When a student draws a Kaboom Card (BLM), he returns all collected cards to the pile.

The player with the most cards at the end of the allotted time wins.

Chapter 6

Division

Find the quotients.
What patterns do you see?

2 ÷ 2	3 ÷ 3	4 ÷ 4	5 ÷ 5	10 ÷ 10
4 ÷ 2	6 ÷ 3	8 ÷ 4	10 ÷ 5	20 ÷ 10
6 ÷ 2	9 ÷ 3	12 ÷ 4	15 ÷ 5	30 ÷ 10
8 ÷ 2	12 ÷ 3	16 ÷ 4	20 ÷ 5	40 ÷ 10
10 ÷ 2	15 ÷ 3	20 ÷ 4	25 ÷ 5	50 ÷ 10
12 ÷ 2	18 ÷ 3	24 ÷ 4	30 ÷ 5	60 ÷ 10
14 ÷ 2	21 ÷ 3	28 ÷ 4	35 ÷ 5	70 ÷ 10
16 ÷ 2	24 ÷ 3	32 ÷ 4	40 ÷ 5	80 ÷ 10
18 ÷ 2	27 ÷ 3	36 ÷ 4	45 ÷ 5	90 ÷ 10
20 ÷ 2	30 ÷ 3	40 ÷ 4	50 ÷ 5	100 ÷ 10

Make flash cards for the facts you need to practice.

24 ÷ 4	6
front	back

171

171

Lesson 1 Dividing Tens and Hundreds

Objectives

- Divide tens or hundreds by 1 one-digit number when there is no remainder.
- Become familiar with the long division symbol and relate the positions of the dividend, divisor, and quotient to the equation form.

Lesson Materials

- Place-value discs

Think

Provide students with place-value discs and have them work the **Think** problem independently.

Have students write an expression for each item and share how they found their answers.

Learn

Work through the **Think** problem with students as demonstrated in **Learn**. Have students work along with place-value discs as the steps are modeled.

Introduce the long division symbol:

- Beginning with the boxes of diapers, ask students what number in the **Think** problem represents the whole, or the amount we are dividing. Show them where that number goes (below the line in the long division symbol).
- Ask the students how many containers, or equal parts, the items will be shared between. Show students where that number goes (to the left of the curved or straight line).
- Ask students how many items will be in each container. Then show them where the quotient goes (above the line).

Tell students that the written equation in this format is often referred to as "long division."

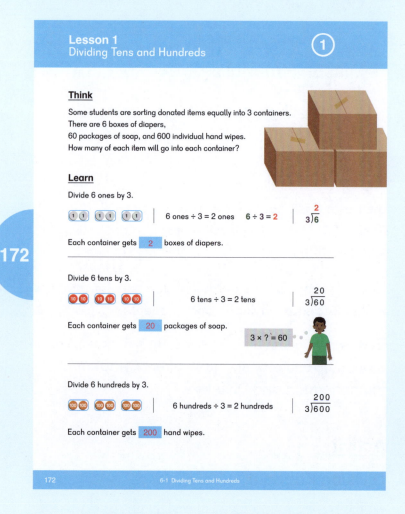

172

Avoid using non-mathematical language to describe the long division symbol, such as, "the big number goes inside the house."

Note that when we divide 6 tens, there are no ones to divide and we can simply write 0 in the ones place in the quotient. Emphasize place value. "When we are dividing the tens, that number of tens in the quotient is written in the tens column."

Repeat for the soap and hand wipes.

After students have worked the problems with place-value discs, have them compare their methods from **Think** with the method shown in the textbook.

Do

Allow students to work the problems with place-value discs if needed. Most should not need manipulatives.

4 Have students rewrite the expressions in the long division format. Check that students are putting the dividend, divisor, and quotient in the correct places and are aligning the digits correctly.

Exercise 1 • page 157

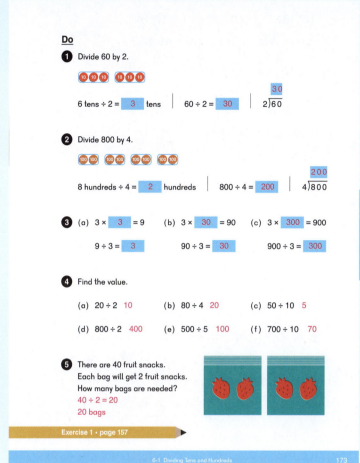

Lesson 2 Dividing a 2-Digit Number by 2 — Part 1

Objective

- Divide a number of up to two digits by 2 when there is a remainder.

Lesson Materials

- Base ten blocks (tens and ones)
- Place-value discs

Think

Provide students with place-value discs and have them work the **Think** problem independently.

Ask students:

- How is this problem different from the ones you solved in the previous lesson? (The numbers don't divide evenly. There is one left over.)
- How is it the same? (We are still dividing by 2.)

Have students write an expression for each item and share how they found their answers.

Learn

Using base ten blocks or place-value discs first, work through the **Think** problems as demonstrated in **Learn**.

Have students work along with manipulatives as the steps are being modeled.

Use the textbook explanations (in green boxes) for suggested language and discuss where each number belongs in the algorithm. For example:

- There are 9 bottles of shampoo. That is the whole that I am dividing. Where is the 9 written in a long division problem?
- The bottles are put into 2 groups. Where do we write the 2 to represent 2 groups?
- When we divide the 9 bottles into 2 groups, each group will have 4 bottles. Where do we place the quotient in the problem?
- Show students where the remainder is written.

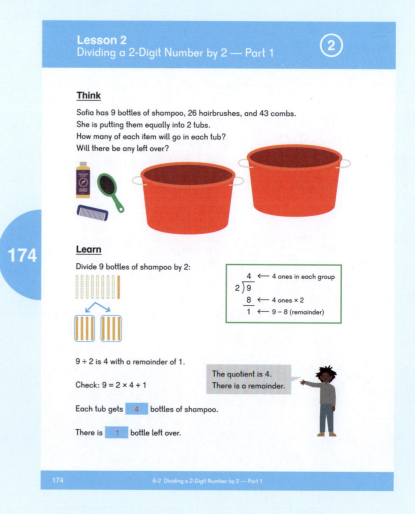

Ask students to recall how to write the answer to a division problem when there is a remainder.

$9 \div 2$ is 4 with a remainder of 1, or 4 R 1. Students can check their work by finding that $9 = 2 \times 4 + 1$.

Ask students what is different about the hair brush problem. They should note that the quotient is a two-digit number.

Ask similar questions about writing the long division problem:

- Where is the whole written in a long division problem?
- Where do we write how many groups we are making or the number of groups we are dividing into?
- How many tens are in 26?
- Where do we record the number of tens in the quotient?
- Where do we record the number of ones in each group in the quotient?
- Are there any ones (hair brushes) left over?

After the students have worked the problems with place-value discs, have them compare their methods from **Think** with the method shown in the textbook.

Mei sees that the tens and ones can be split and then divided separately.

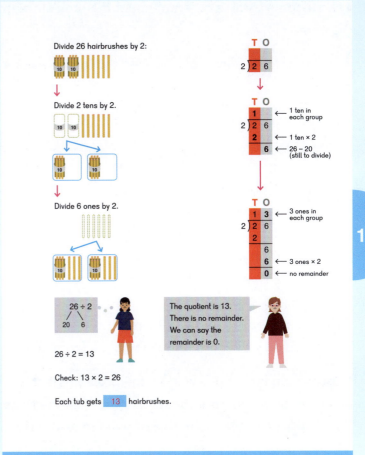

Ask students what is different about the problem with the combs.

Ask similar questions about writing the long division problem:

- Where is the whole written in a long division problem?
- Where do we write how many groups we are making or the number of groups we are dividing into?
- How many tens are in 43?
- Where do we record the number of tens in each group in the quotient?
- Where do we record the number of ones in each group in the quotient?
- Are there any (combs) left over?
- Where do we record remaining ones (combs)?

After the students have worked the problems with place-value discs, have them compare their methods from **Think** with the method shown in the textbook.

Have students check their division work with multiplication.

Alex thinks 40 ÷ 2 is 20 and 3 ÷ 2 is 1 remainder 1, or 21 remainder 1.

Do

❸ Provide practice by having students work these problems with base ten blocks or place-value discs.

Have students rewrite the expressions in the long division format. Check that students are putting the dividend, divisor, and quotient in the correct places and are aligning the digits correctly.

Exercise 2 • page 159

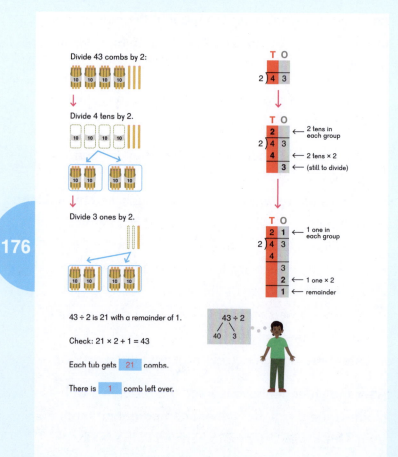

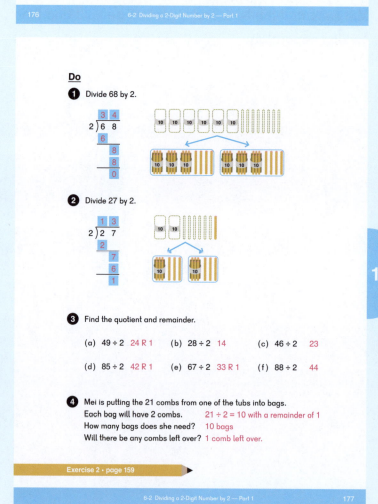

Do

❶ Divide 68 by 2.

❷ Divide 27 by 2.

❸ Find the quotient and remainder.

(a) 49 ÷ 2　24 R 1　　(b) 28 ÷ 2　14　　(c) 46 ÷ 2　23

(d) 85 ÷ 2　42 R 1　　(e) 67 ÷ 2　33 R 1　　(f) 88 ÷ 2　44

❹ Mei is putting the 21 combs from one of the tubs into bags.
Each bag will have 2 combs.　21 ÷ 2 = 10 with a remainder of 1
How many bags does she need?　10 bags
Will there be any combs left over?　1 comb left over.

Exercise 2 • page 159

Lesson 3 Dividing a 2-Digit Number by 2 — Part 2

Objective

- Divide a two-digit number by 2 when there is a remainder for tens.

Lesson Materials

- Place-value discs
- Base ten blocks (tens and ones)

Think

Provide students with place-value discs and have them work the **Think** problems independently.

Ask students:

- How are these problem different from the ones you solved in the previous lesson? (We have to regroup the tens to divide them into two equal groups.)
- How is it the same? (We can still divide the digit in each place.)

Have students write an expression for each item and share how they found their answers.

Learn

Using place-value discs, work through the **Think** problems as demonstrated in **Learn**.

Have students work along with place-value discs as the steps are modeled.

The toothpaste tube problem is fairly straightforward, as students should already know the answer to 14 ÷ 2. Focus on the fact that since we cannot divide 1 ten into two groups of ten, we have to regroup the ten as 10 ones, so we have 14 ones, which can be divided into more than one group.

Discuss:

- There are 14 tubes of toothpaste to divide into 2 groups. Where does 14 go in the long division algorithm?

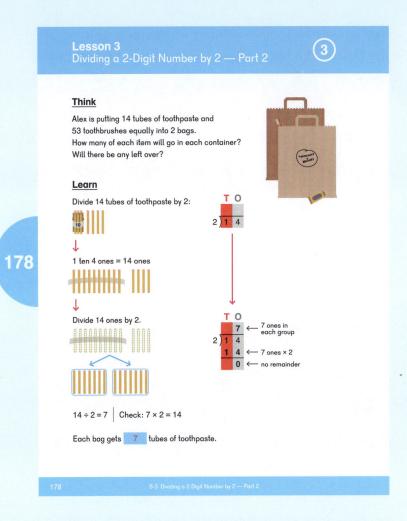

- The tubes are put into 2 groups. Where do we write the digit 2 to represent 2 groups?

One 10-disc can't be shared equally among two groups. Have students regroup one 10-disc as ten 1-discs. There are now 14 ones.

- When we divide the 14 tubes into 2 groups, each group will have 7 tubes. How do we show this with place-value discs?
- Where do we place the quotient in the problem?

Discussing 53 toothbrushes ÷ 2 groups may take the bulk of the lesson time. Do not rush through this problem.

- Where does 53 go in the long division algorithm?
- Where do we write the 2 to represent 2 groups?

Divide the 5 tens into 2 groups. There are now 2 tens in each group with 1 ten remaining to be divided.

- How do we record the tens that we divided and the ten remaining in our long division algorithm?
- What did we do in our first problem when we had 1 ten to divide?

Regroup the remaining 1 ten as 10 ones. There are now 13 ones.

- How are the 13 ones recorded in the long division algorithm?

Divide the 13 into 2 groups. Each group will have 6 and there is 1 remaining.

- How do we show this with place-value discs?
- How do we write this in the long division format?

After the students have worked the problems with place-value discs, have them compare their methods from **Think** with the method shown in the textbook.

Sofia splits 53 into 40 and 13 because she can easily divide 40 by 2 and 13 by 2.

Students can use this idea of splitting the divisor into two numbers that they can divide easily by 2 to solve some problems mentally.

Note: Students adept at mental math may also split 53 into 50 and 3, knowing that 50 ÷ 2 = 25.

Students should be encouraged to use mental strategies when possible.

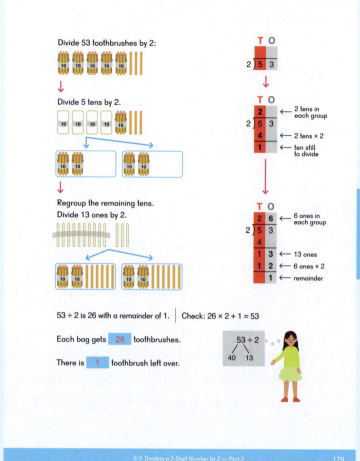

Divide 53 toothbrushes by 2:

Divide 5 tens by 2.

Regroup the remaining tens.
Divide 13 ones by 2.

2 tens in each group
2 tens × 2
ten still to divide

6 ones in each group
13 ones
6 ones × 2
remainder

53 ÷ 2 is 26 with a remainder of 1. | Check: 26 × 2 + 1 = 53

Each bag gets 26 toothbrushes.

There is 1 toothbrush left over.

53 ÷ 2
40   13

6-3 Dividing a 2-Digit Number by 2 — Part 2   179

179

Do

①—② Ensure students are relating the steps with place-value discs to the written process in the problem.

②—③ Mei and Emma find easier numbers to use a mental math strategy to divide.

After students are proficient with the algorithm, some will be able to use this idea to solve some problems mentally rather than writing out the long division. They do have to mentally add the quotients, 30 + 7, and remember there is a remainder.

④ Have students practice these problems with base ten blocks or place-value discs, as needed.

Have students rewrite the expressions in the long division format. Check that students are putting the dividend, divisor, and quotient in the correct places and are aligning the digits correctly.

Students will continue to practice the steps in the written algorithm in the next lesson.

⑤ This problem elaborates on the relationship between odd numbers and remainders. Ask students, "When dividing by 2, why is it not possible to have a remainder greater than 1?"

Exercise 3 • page 161

Lesson 4 Dividing a 2-Digit Number by 3, 4, and 5

Objective

- Divide a two-digit number by 3, 4, and 5.

Lesson Materials

- Place-value discs

Think

Provide students with place-value discs and have them work the **Think** problem independently.

Ask students:

- How is this problem different from the ones you solved in the previous lesson? (We are dividing by 3.)
- How is it the same? (We can still divide the digit in each place.)
- Will there be a remainder? (Students adept at mental math may realize that 25 × 3 = 75, so there will be a remainder.)

Have students write expressions for each item and share how they found their answers.

Learn

Using place-value discs first, work through the **Think** problems as demonstrated in **Learn**.

Have students work along with place-value discs as the steps are modeled.

Note that both problems will use the same method to divide.

Discuss:

(a) • There are 76 bars of soap to divide into 3 groups. Where does 76 go in the long division algorithm? Where do we write the 3 to represent 3 groups?

Divide the tens into 3 groups. There are now 2 tens in each group with 1 ten remaining to be divided.

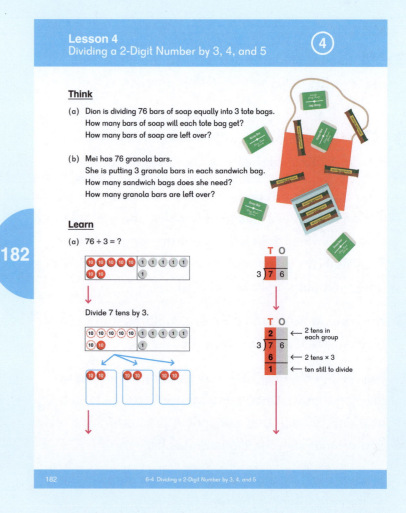

Think

(a) Dion is dividing 76 bars of soap equally into 3 tote bags. How many bars of soap will each tote bag get? How many bars of soap are left over?

(b) Mei has 76 granola bars. She is putting 3 granola bars in each sandwich bag. How many sandwich bags does she need? How many granola bars are left over?

Learn

(a) 76 ÷ 3 = ?

Divide 7 tens by 3.

182   6-4 Dividing a 2-Digit Number by 3, 4, and 5

- How do we record the tens in each group and the 1 ten remaining in our long division algorithm?

Regroup the remaining one 10-disc as ten 1-discs. There are now 16 ones.

- How are the 16 ones recorded in the long division algorithm?

Divide 16 bars of soap into 3 groups. Each group will have 5 bars and there is 1 bar of soap remaining.

- How do we show this with place-value discs?
- How do we write this in the algorithm?

(b) Have students compare the phrasing in (a) to the phrasing in (b). Problem (a) is sharing division. We are sharing the soap among 3 containers. Problem (b) is grouping division. In this problem, the number of groups is the quotient, not the divisor.

Note: Grouping division situations are not shown with place-value discs, yet the algorithm is the same. For 60 granola bars, 3 to a bag, we need 20 bags. For 16 granola bars with 3 to a bag, we need an additional 5 bags and there is 1 granola bar left over.

Dion uses a mental strategy by thinking of the greatest number of tens (6 tens) that divides evenly by 3.

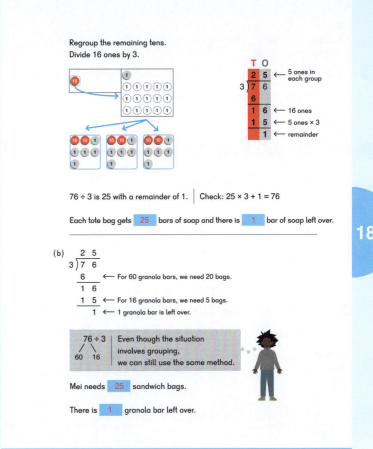

Do

These problems have 3, 4, and 5 as the divisor.

①—② Ensure students are relating the steps with place-value discs to the written process in the problem.

① Discuss Alex's thought on the amount of the remainder when dividing by 4. Pose the same question for dividing by 3 and 5.

④ Provide practice by having students work these problems with place-value discs, as needed.

Have students rewrite the expressions in the long division format. Check that students are putting the dividend, divisor, and quotient in the correct places and are aligning the digits correctly.

Exercise 4 • page 164

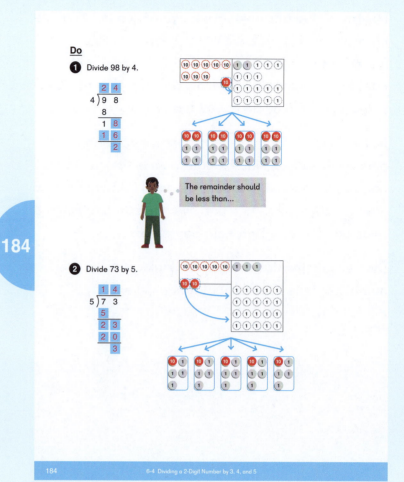

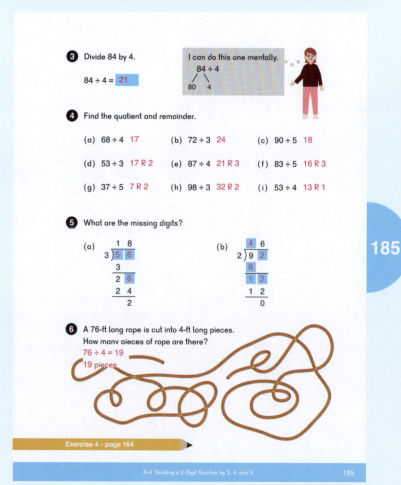

Objective

- Practice division.

Sample bar models:

5

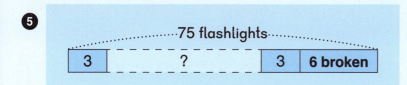

```
·············75 flashlights·············
┌─────┬─────────────┬─────┬──────────┐
│  3  │      ?      │  3  │ 6 broken │
└─────┴─────────────┴─────┴──────────┘
```

6

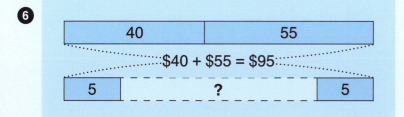

```
┌──────────────┬──────────────┐
│      40      │      55      │
└──────────────┴──────────────┘
····$40 + $55 = $95····
┌─────┬──────────────┬─────┐
│  5  │      ?       │  5  │
└─────┴──────────────┴─────┘
```

7 If we remove 28 from the top bar there will be equal units. That is, 2 units + 28 = 100 or 100 − 28 = 72.

2 units = 72

1 unit = 72 ÷ 2, or 36

8 Alex provides thinking on how to access this problem:

- Subtract the difference between Adam and Chapa's amounts saved, or $7 from the total: $92 − $7 = $85
- Subtract the difference between Adam and Ethan: $12 + $7, or $19 from the remaining amount: $85 − $19 = $66.
- Divide $66 into 3 equal units: 66 ÷ 3 = 22.

Adam saved $22.
Chapa saved $22 + $7, or $29.
Ethan saved $29 + $12 or $41.

---

1 Find the quotient.

(a) 600 ÷ 2 300   (b) 90 ÷ 3 30   (c) 800 ÷ 4 200

(d) 80 ÷ 2 40   (e) 500 ÷ 5 100   (f) 200 ÷ 2 100

2 Find the quotient and remainder.

(a) 46 ÷ 2 23   (b) 81 ÷ 3 27   (c) 79 ÷ 4 19 R 3

(d) 65 ÷ 5 13   (e) 88 ÷ 3 29 R 1   (f) 69 ÷ 2 34 R 1

(g) 72 ÷ 4 18   (h) 27 ÷ 2 13 R 1   (i) 99 ÷ 5 19 R 4

(j) 77 ÷ 3 25 R 2   (k) 93 ÷ 5 18 R 3   (l) 83 ÷ 2 41 R 1

3 Wainani is putting 75 rolls of paper towels equally into 4 boxes. 75 ÷ 4 is 18 R 3
How many rolls of paper towels will be in each box? 18 rolls
How many will be left over? 3 rolls
Check: 4 × 18 + 3 = 75

4 There are 89 people going on the Ferris wheel.
One Ferris wheel car can hold 3 people.
Each car is filled before people get into the next car.
How many Ferris wheel cars have people in them? 89 ÷ 3 is 29 R 2
30 cars have people in them.

5 Hunter has 75 flashlights to put into boxes.
Each box will get 3 flashlights.
He first tested each and found that 6 did not work and discarded them.
How many boxes did he use? 75 − 6 = 69
69 ÷ 3 = 23
23 boxes

6 Aurora saved $40 last month and $55 this month.
She wants to buy gifts that cost $5 each for her friends.
How many gifts can she buy? 40 + 55 = 95
95 ÷ 5 = 19
19 gifts

7 The sum of two numbers is 100.
The difference between the two numbers is 28.
What are the two numbers?

100 − 28 = 72
72 ÷ 2 = 36
36 + 28 = 64
Check: 36 + 64 = 100
The numbers are 36 and 64.

```
100
28
```

8 Adam, Chapa, and Ethan saved a total of $92. 66 ÷ 3 = 22
Chapa saved $7 more than Adam.
Ethan saved $12 more than Chapa.

Adam saved $22.

92 − 12 − 7 − 7 = 66

If I make Adam's bar 1 unit, then I can subtract 7, 12, and 7 to have 3 equal units.

(a) How much money did Adam save?

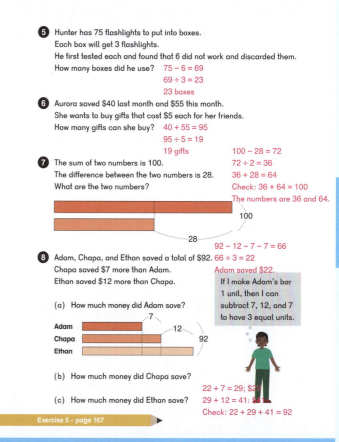

```
        7
Adam  ───┐
Chapa ────┐ 12      92
Ethan ─────┘
```

(b) How much money did Chapa save?

22 + 7 = 29; $29

(c) How much money did Ethan save?

29 + 12 = 41; $41
Check: 22 + 29 + 41 = 92

Exercise 5 • page 167

Activities

▲ **Leftovers**

Materials: 45 counters, die with modified sides: 0 in place of 6

Extend this game, first introduced in Chapter 4: Lesson 8, on page 149 of this Teacher's Guide, to include writing the long division problem.

Players take turns rolling the die and dividing the counters by the number on the die.

```
      2   2
   2 ) 4   5
      4
      ‾‾‾
          5
          4
          ‾‾‾
          1
```

For example, Player One rolls a 2. He divides the counters into 2 equal groups with 1 left over. He then keeps the leftover counter and play continues with the remaining 44 counters.

```
          8
   5 ) 4   4
      4   0
      ‾‾‾‾‾
          4
```

Player Two rolls a 5 and divides the remaining 44 counters by 5. She has 8 groups of 5, with 4 counters leftover. She then keeps the 4 counters and returns the 40 remaining counters.

When all the counters have been used, the game is over. The player with the most counters wins.

▲ **Grab and Divide**

Materials: Tens and ones place-value discs in a bag, die with modified sides: "roll again" in place of 6, recording sheet

In each round, players pass the bag and take 10 place-value discs without looking. Players use place-value discs to make a two-digit number.

Players then take turns rolling the die and dividing their two-digit number by the number shown on the die.

The player with the greatest quotient is the winner. In case of a tie, the player with the greatest quotient plus remainder wins.

◀ **Exercise 5 • page 167**

Lesson 6 Dividing a 3-Digit Number by 2

Objective

- Divide three-digit numbers by 2 when there is regrouping in the tens and/or ones.

Lesson Materials

- Place-value discs

Think

Provide students with place-value discs and have them work the **Think** problem independently.

Ask students:

- How is this problem different from the ones you solved in the previous lesson? (We are dividing three-digit numbers.)
- How is it the same? (We can still divide the digit in each place.)
- Will there be a remainder?

Have students write expressions and share how they found their answers.

Learn

Using place-value discs first, work through the **Think** problems as demonstrated in **Learn**.

Have students work along with place-value discs as the steps are modeled.

Discuss:

- There are 531 scarves to divide into 2 groups. Where does 531 go in the long division algorithm? Where do we write the 2 to represent 2 groups?

Divide the hundreds into 2 groups. There are now 2 hundreds in each group with 1 hundred remaining to be divided.

- How do we record the hundreds in each group and the 1 hundred remaining in our long division algorithm?

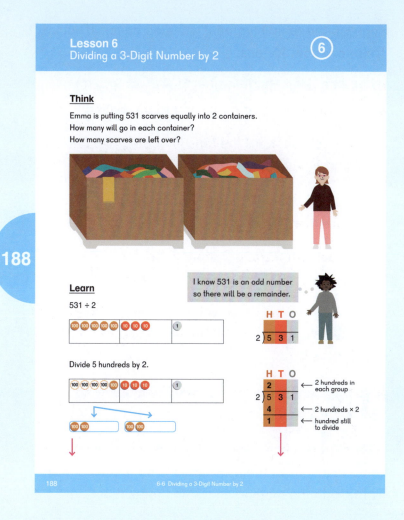

Regroup the remaining one 100-disc as ten 10-discs. There are now 13 tens.

Divide the tens into 2 groups. There are now 6 tens in each group with 1 ten remaining to be divided.

- How do we record the tens in each group and the 1 ten remaining in our long division algorithm?

Regroup the remaining one 10-disc as ten 1-discs. There are now 11 ones.

Divide the 11 ones into 2 groups. Each group will have 5 ones and there is one 1-disc remaining.

- How do we show this with place-value discs? How do we write this in the long division problem?

Students can check their answer by multiplying: 265 × 2 + 1 = 531

After students have worked the problems with place-value discs, have them compare their methods from **Think** with the method shown in the textbook.

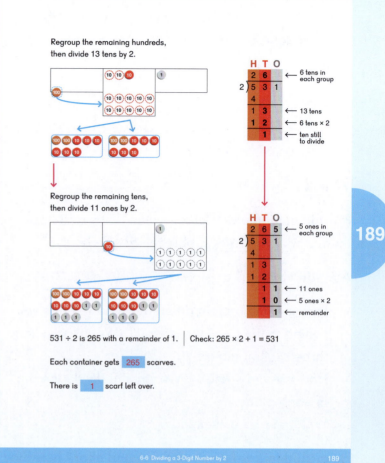

Regroup the remaining hundreds, then divide 13 tens by 2.

Regroup the remaining tens, then divide 11 ones by 2.

531 ÷ 2 is 265 with a remainder of 1. | Check: 265 × 2 + 1 = 531

Each container gets 265 scarves.

There is 1 scarf left over.

6-6 Dividing a 3-Digit Number by 2   189

Do

1–**2** Ensure students are relating the steps with place-value discs to the written process in the problem.

Help students see the connection between **1** and **2**.

750 is 50 more than 700, and 50 ÷ 2 is 25, so 750 ÷ 2 is 25 more than 700 ÷ 2.

4 Sofia thinks that this is an easy problem to divide mentally. She sees that all the digits are even, so there is no regrouping. She splits 864 into 8 hundreds, 6 tens, and 4 ones.

8 hundreds ÷ 2 = 4 hundreds

6 tens ÷ 2 = 3 tens

4 ones ÷ 2 = 2 ones

5 Have students practice by working these problems with place-value discs.

Have students rewrite the expressions in the long division format. Check that students are putting the dividend, divisor, and quotient in the correct places and are aligning the digits correctly.

Exercise 6 • page 170

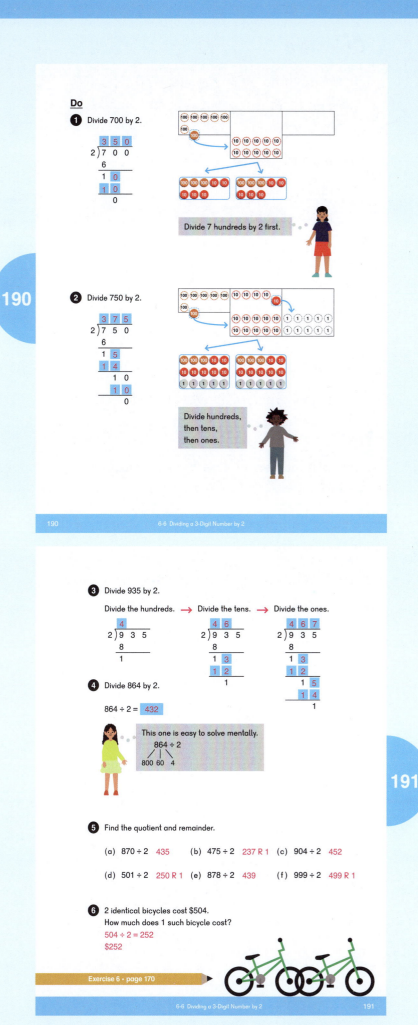

Lesson 7 Dividing a 3-Digit Number by 3, 4, and 5

Objective

- Divide three-digit numbers by 3, 4, and 5 when there is regrouping in the hundreds, tens, and ones.

Lesson Materials

- Place-value discs

Think

Provide students with place-value discs and have them work the **Think** problem independently.

Ask students:

- How is this problem different from the ones you solved in the previous lesson? (We are dividing by 3.)
- How is it the same? (We can still divide the digit in each place.)

Have students write expressions and share how they found their answers.

Discuss Emma's thought about remainders. There should never be a remainder in any column equal to or greater than the divisor.

Learn

Using place-value discs first, work through the **Think** problems as demonstrated in **Learn**.

Have students work along with place-value discs as the steps are modeled.

Discuss the regrouping with place-value discs and show the written steps:

- Divide 4 hundreds by 3.
- Regroup the remaining 1 hundred into 10 tens.

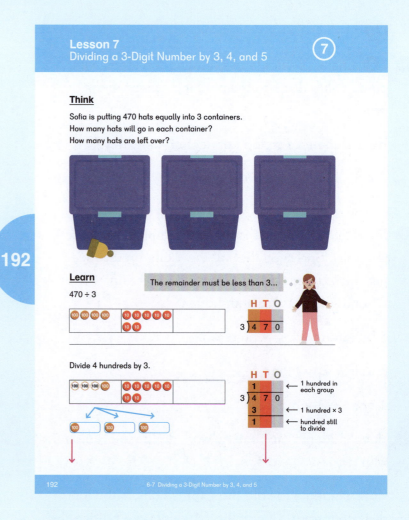

- Divide 17 tens by 3.
- Regroup the remaining 2 tens into 20 ones.
- Divide 20 ones by 3. Each group has 6 with a remainder of 2.

Students can check their answer, "470 ÷ 3 = 156 remainder 2," with multiplication: 156 × 3 + 2.

After the students have worked the problems with place-value discs, have them compare their methods from **Think** with the method shown in the textbook.

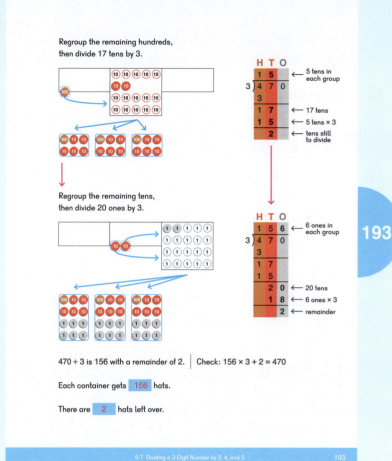

Regroup the remaining hundreds, then divide 17 tens by 3.

H	T	O	
1	5		← 5 tens in each group
3) 4	7	0	
3			
1	7		← 17 tens
1	5		← 5 tens × 3
	2		← tens still to divide

Regroup the remaining tens, then divide 20 ones by 3.

H	T	O	
1	5	6	← 6 ones in each group
3) 4	7	0	
3			
1	7		
1	5		
	2	0	← 20 tens
	1	8	← 6 ones × 3
		2	← remainder

470 ÷ 3 is 156 with a remainder of 2. | Check: 156 × 3 + 2 = 470

Each container gets 156 hats.

There are 2 hats left over.

193

Do

Students who can do these problems without place-value discs should be encouraged to do so.

1 Alex points out that students will regroup 2 tens for 20 ones. They must record a 0 in the tens place in the quotient.

3 Note that there are no tens to add to the regrouped 20 tens. Students who are struggling should solve the problem with place-value discs first, then relate their steps to the picture in the textbook.

4 Sofia thinks that this is an easy problem to divide mentally. She sees that each digit in 693 divides evenly by 3. She splits 639 into 6 hundreds, 3 tens, and 9 ones.

6 hundreds ÷ 3 = 2 hundreds

3 tens ÷ 3 = 1 ten

9 ones ÷ 3 = 3 ones

5 Have students rewrite the expressions in the long division format. Check that students are putting the dividend, divisor, and quotient in the correct places and are aligning the digits correctly. Have students share mental strategies for some of these problems.

Exercise 7 • page 174

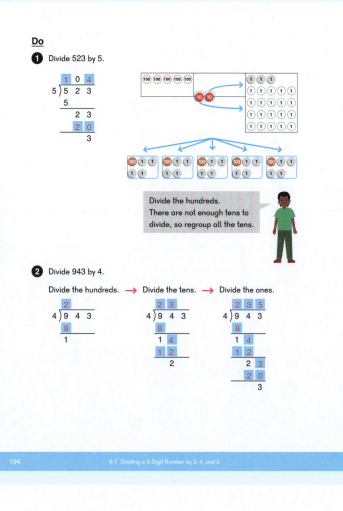

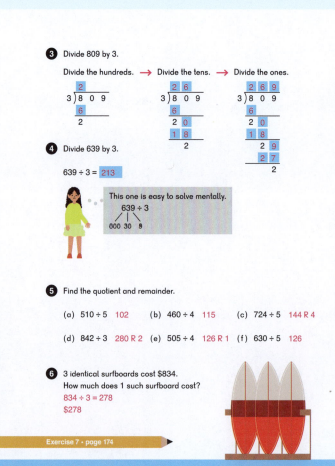

Lesson 8 Dividing a 3-Digit Number, Quotient is 2 Digits

Objective

- Divide a three-digit number when the digit in the hundreds place cannot be divided by the divisor.

Lesson Materials

- Place-value discs

Think

Provide students with place-value discs and have them work the **Think** problem independently.

Ask students:

- How is this problem different from the ones you solved in the previous lesson? (We don't have enough hundreds for 4 equal groups.)
- How is it the same? (We can still divide the digit in each place.)
- Will there be a remainder?

Have students write expressions and share how they found their answers.

Learn

Using place-value discs first, work through the **Think** problems as demonstrated in **Learn**.

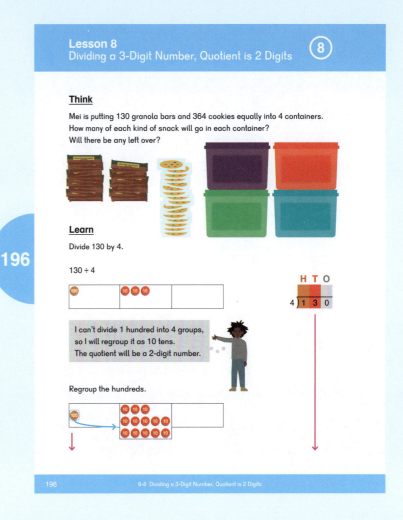

Have students work along with place-value discs as the steps are modeled.

Discuss the regrouping with place-value discs and show the written steps:

- Regroup 1 hundred into 10 tens. There are now 13 tens.
- Divide 13 tens by 4.
- Regroup the remaining 1 ten into 10 ones. There are now 10 ones.
- Divide 10 ones by 4.

Students can check the answer, "130 ÷ 4 is 32 with a remainder of 2," with multiplication: 32 × 4 + 2 = 130.

After the students have worked the problems with place-value discs, have them compare their methods from **Think** with the method shown in the textbook.

Discuss Alex's mental method for 364 ÷ 4. He knows 360 or 36 tens ÷ 4 is 90, and 4 ÷ 4 is 1, and 90 + 1 = 91.

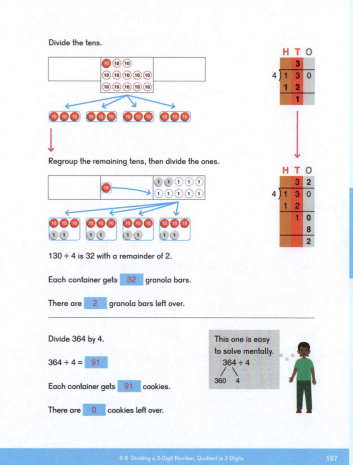

197

Do

2 Students who are struggling should solve the problem with place-value discs first, then relate the steps they did with place-value discs to the picture in the textbook.

3 Students should see that when the number in the hundreds place that is to be divided is less than the number of groups, the quotient will be a two-digit number.

4

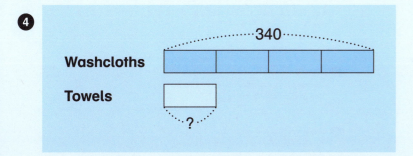

Washcloths

Towels

340

?

Exercise 8 • page 177

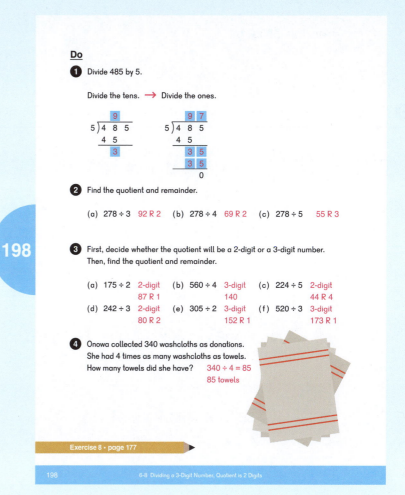

Lesson 9 Practice B

Objective

- Practice division.

1 Possible mental strategies:

(a)

$$64 \div 2 = 30 + 2 = 32$$

60   4

(b)

$$903 \div 3 = 300 + 1 = 301$$

900   3

(c)

$$364 \div 4 = 90 + 1 = 91$$

360   4

(d)

$$80 \div 4 = 10 + 10 = 20$$

40   40

(e)

$$824 \div 4 = 200 + 6 = 206$$

800   24

(f)

$$255 \div 5 = 50 + 1 = 51$$

250   5

Lesson 9   P **9**
Practice B

1 Find the quotient.
Try to solve mentally.

(a) $64 \div 2$ 32   (b) $903 \div 3$ 301   (c) $364 \div 4$ 91

(d) $80 \div 5$ 16   (e) $824 \div 4$ 206   (f) $255 \div 5$ 51

2 Find the quotient and remainder.

(a) $900 \div 2$ 450   (b) $809 \div 2$ 404 R 1   (c) $297 \div 3$ 99

(d) $192 \div 4$ 48   (e) $242 \div 3$ 80 R 2   (f) $197 \div 2$ 98 R 1

(g) $345 \div 4$ 86 R 1   (h) $787 \div 3$ 262 R 1   (i) $167 \div 3$ 55 R 2

(j) $291 \div 4$ 72 R 3   (k) $459 \div 5$ 91 R 4   (l) $409 \div 5$ 81 R 4

3 A baseball coach has $187.
He wants to buy baseballs that cost $5 each.
How many baseballs can he buy? $187 \div 5$ is 37 R 2
37 baseballs

4 Jason has 210 m of ribbon.
He wants to cut it into pieces each 4 m long.
How many 4 m long pieces will he have?
$210 \div 4$ is 52 R 2
52 pieces

Baseballs

6-9 Practice B   199

199

5–**8** Students may find it helpful to draw bar models.

8 Note that the two bars make equal units up to the difference of 90. That is, 2 units + 90 = 560 or 560 − 90 = 470.

>   2 units = 470
>
>   1 unit = 470 ÷ 2, or 235

9 Sofia's thoughts provide a hint. This becomes an easier problem if there are equal units. She can make 2 equal units by adding 20 to Pablo's amount.

Altogether, $155 is the sum of Sara's amount + Anna's amount + Pablo's amount: 155 = 1 unit + 2 units + 2 units − 20. If $20 is added to Pablo's amount, 5 units = $175.

1 unit is $175 ÷ 5 = $35.

Sara has $35.

Anna has 35 × 2 = $70.

Pablo has $70 − $20 = $50.

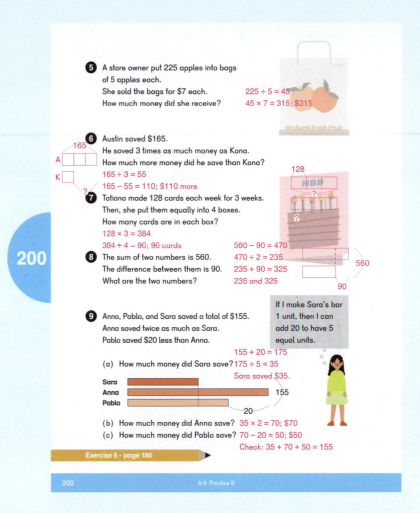

Activity

▲ Divvy Up

Materials: Divvy Up Game Cards (BLM), 4 sets of Number Cards (BLM) 0 to 9

Players take turns drawing three Number Cards (BLM) and creating a three-digit number. They write their numbers on one of the four division problem lines on a Divvy Up Game Card (BLM).

Players then complete the division problem recording the quotient and remainder (if any). After four rounds, players total up their quotients and remainders and add them together.

The player with the greatest total score is the winner.

Alternatively, the player with the least total score wins.

◀ **Exercise 9 • page 180**

Brain Works

★ If I Had a Million Dollars...

Your parents offer you a special allowance for one month.

Option 1: Each day that month you get double the number of pennies you received on the previous day.

Option 2: You can have a million dollars.

Which do you choose?

Chapter 6 Division

Exercise 1

Basics

1 (a) 8 ones ÷ 4 = **2** ones

$= $ **2**

$4\overline{)8}$ → **2**

(b) 8 tens ÷ 4 = **2** tens

$= $ **20**

$4\overline{)80}$ → **20**

(c) 8 hundreds ÷ 4 = **2** hundreds

$= $ **200**

$4\overline{)800}$ → **200**

2 (a) 12 ones ÷ 3 = **4** ones

$= $ **4**

$3\overline{)12}$ → **4**

(b) 12 tens ÷ 3 = **4** tens

$= $ **40**

$3\overline{)120}$ → **40**

(c) 12 hundreds ÷ 3 = **4** hundreds

$= $ **400**

$3\overline{)1200}$ → **400**

Practice

3 (a) 4 hundreds ÷ 2 = **2** hundreds = **200**

(b) 15 tens ÷ 3 = **5** tens = **50**

(c) 20 hundreds ÷ 4 = **5** hundreds = **500**

(d) 25 tens ÷ 5 = **5** tens = **50**

4 Divide.

$3\overline{)9}$ → **3**	$3\overline{)90}$ → **30**	$3\overline{)900}$ → **300**
$4\overline{)16}$ → **4**	$4\overline{)160}$ → **40**	$4\overline{)1600}$ → **400**
$5\overline{)10}$ → **2**	$5\overline{)100}$ → **20**	$5\overline{)1000}$ → **200**

Challenge

5 (a) 40 tens ÷ 2 = **200**

(b) 20 tens ÷ 5 = **40**

6 Find the sum of 40 tens and 2 hundreds divided by 3.

400 + 200 = 600

600 ÷ 3 = 200

Exercise 2

Basics

1 Fill in the missing digits and numbers.

(a) Divide 65 by 2.

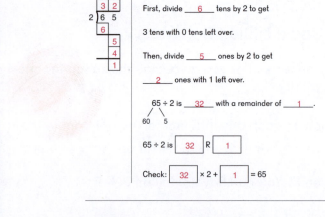

First, divide ___**6**___ tens by 2 to get

3 tens with 0 tens left over.

Then, divide ___**5**___ ones by 2 to get

___**2**___ ones with 1 left over.

65 ÷ 2 is ___**32**___ with a remainder of ___**1**___.

60   5

65 ÷ 2 is **32** R **1**

Check: **32** × 2 + **1** = 65

(b) Divide 48 by 2.

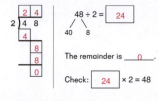

48 ÷ 2 = **24**

40   8

The remainder is ___**0**___.

Check: **24** × 2 = 48

Practice

2 Find the quotient and remainder.

$2\overline{)28}$ → **14** Quotient **14**; Remainder **0**

$2\overline{)84}$ → **42** Quotient **42**; Remainder **0**

$2\overline{)69}$ → **34** Quotient **34**; Remainder **1**

$2\overline{)43}$ → **21** Quotient **21**; Remainder **1**

$2\overline{)81}$ → **40** Quotient **40**; Remainder **1**

$2\overline{)25}$ → **12** Quotient **12**; Remainder **1**

3 A paddle boat holds 2 people.

How many paddle boats are needed for 27 people?

14 paddle boats

Exercise 3

Basics

1 Fill in the missing digits and numbers.

(a) Divide 73 by 2.

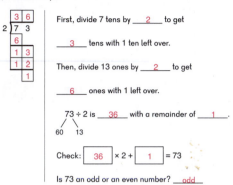

First, divide 7 tens by __2__ to get

__3__ tens with 1 ten left over.

Then, divide 13 ones by __2__ to get

__6__ ones with 1 left over.

73 ÷ 2 is __36__ with a remainder of __1__.

Check: [36] × 2 + [1] = 73

Is 73 an odd or an even number? __odd__

(b) Divide 54 by 2.

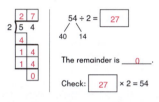

54 ÷ 2 = [27]

The remainder is __0__.

Check: [27] × 2 = 54

Practice

2 Divide.

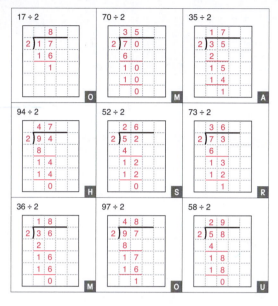

17 ÷ 2	70 ÷ 2	35 ÷ 2
94 ÷ 2	52 ÷ 2	73 ÷ 2
36 ÷ 2	97 ÷ 2	58 ÷ 2

Riddle: What kind of room has no doors or windows?
Write the letters in the boxes below to find out.

A	M	U	S	H	R	O	O	M
48	17 R 1	26 R 1	18	29	26	47	36 R 1	8 R 1	48 R 1	35

3 Megan wants to put 52 cookies evenly into 2 cookie jars.
How many cookies will be in each jar?
52 ÷ 2 = 26
26 cookies will be in each jar.

4 A tailor is sewing 2 buttons on each cuff of a jacket.
He has 58 buttons.
How many jackets can he sew buttons on both cuffs?
2 × 2 = 4
58 ÷ 4 is 14 R 2
He can sew buttons on 14 jackets.

Challenge

5 Mei is holding a card with an even number in one hand, and a card with an odd number in the other hand.
Dion tells her to triple the value of the card in her right hand and double the value of the card in her left hand, and then add the two products.
If the sum is even, which hand is holding the even card?

If the sum is even, both products must be even.
If you triple an odd number, the product is odd.
If you double an odd number, or double or triple an even number, the product is even.
If her right hand held an odd card, the sum would be odd.
Her right hand is holding the even card.

Exercise 4

Basics

1 Fill in the missing digits and numbers.

(a) Divide 83 by 3.

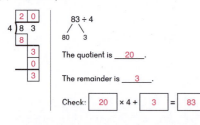

First, divide 8 tens by __3__ to get

__2__ tens with 2 tens left over.

Then, divide 23 ones by __3__ to get

__7__ ones with __2__ left over.

83 ÷ 3 is __27__ with a remainder of __2__.
60   23

Check: [27] × 3 + [2] = 83

(b) Divide 83 by 4.

83 ÷ 4
80   3

The quotient is __20__.

The remainder is __3__.

Check: [20] × 4 + [3] = [83]

Practice

2 Divide.

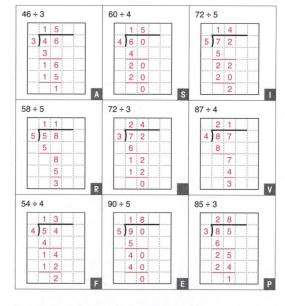

46 ÷ 3	60 ÷ 4	72 ÷ 5
A	S	I
58 ÷ 5	72 ÷ 3	87 ÷ 4
R		V
54 ÷ 4	90 ÷ 5	85 ÷ 3
F	E	P

How many hearts does the common earthworm have?
Write the letters (or blank for space) in the boxes below to find out.

F	I	V	E		P	A	I	R	S	
13 R 2	14 R 2	21 R 3	18	24	28 R 1	15 R 1	14 R 2	11 R 3	15	13 R 3

3 Jamal collected 78 cans of food for a food bank drive.
He collected 3 times as many cans as Landon.
How many cans did Landon collect?
78 ÷ 3 = 26
Landon collected 26 cans.

4 A car can hold 5 people.
What is the fewest number of cars needed for 62 people?
62 ÷ 5 is 12 R 2
The fewest number of cars needed is 13 cars.

Challenge

5 Write the missing digits.

(a)

(b)

Exercise 5

Check

1 Divide.

98 ÷ 3	57 ÷ 4	85 ÷ 5

99 ÷ 5	99 ÷ 3	99 ÷ 4

2 | 83 | ÷ 3 is 27 with a remainder of 2.

3 Circle the even numbers.

⟨98⟩   37   ⟨50⟩   45   ⟨100⟩

4 Max bought 4 identical stools and a small table for $99.
The table cost $23.
How much did each stool cost?
99 − 23 = 76
76 ÷ 4 = 19
Each stool cost $19.

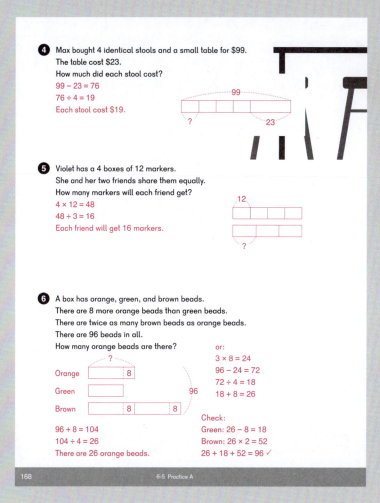

5 Violet has a 4 boxes of 12 markers.
She and her two friends share them equally.
How many markers will each friend get?
4 × 12 = 48
48 ÷ 3 = 16
Each friend will get 16 markers.

6 A box has orange, green, and brown beads.
There are 8 more orange beads than green beads.
There are twice as many brown beads as orange beads.
There are 96 beads in all.
How many orange beads are there?

Orange   8
Green
Brown   8   8

96 + 8 = 104
104 ÷ 4 = 26
There are 26 orange beads.

or:
3 × 8 = 24
96 − 24 = 72
72 ÷ 4 = 18
18 + 8 = 26

Check:
Green: 26 − 8 = 18
Brown: 26 × 2 = 52
26 + 18 + 52 = 96 ✓

Challenge

7 List the odd numbers between 0 and 50 that have a remainder of 4 when
divided by 5.
9, 19, 29, 39, 49

8 Taylor has a box of red and blue beads.
For every red bead, there are two blue beads.
There are 96 beads in all.
How many of the beads are blue?
Make groups of red and blue beads.
96 ÷ 3 = 32 groups of 1 red and 2 blue
2 × 32 = 64
64 of the beads are blue.

9 A box has pink, purple, and yellow beads.
There are 15 pink beads, more purple than pink beads, and more yellow
than purple beads.
There are 96 beads in all.
The difference between the number of pink and purple beads is the
same as the difference between the number of purple and yellow beads.
How many beads are yellow?

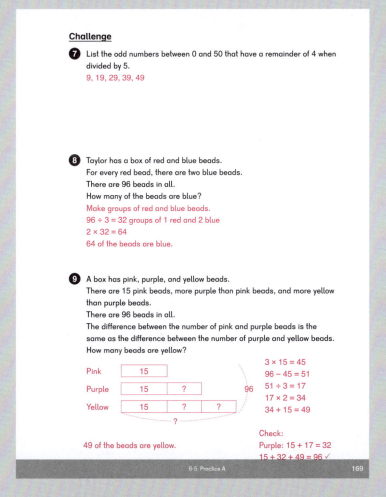

Pink   15
Purple 15   ?   96
Yellow 15   ?   ?

3 × 15 = 45
96 − 45 = 51
51 ÷ 3 = 17
17 × 2 = 34
34 + 15 = 49

Check:
Purple: 15 + 17 = 32
15 + 32 + 49 = 96 ✓

49 of the beads are yellow.

Exercise 6

Basics

1 Fill in the missing digits and numbers.

Divide 753 by 2.

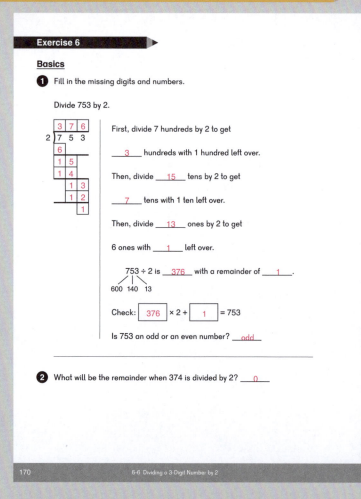

First, divide 7 hundreds by 2 to get

___3___ hundreds with 1 hundred left over.

Then, divide ___15___ tens by 2 to get

___7___ tens with 1 ten left over.

Then, divide ___13___ ones by 2 to get

6 ones with ___1___ left over.

753 ÷ 2 is ___376___ with a remainder of ___1___.

600 140 13

Check: [376] × 2 + [1] = 753

Is 753 an odd or an even number? ___odd___

2 What will be the remainder when 374 is divided by 2? ___0___

Practice

3 Divide.

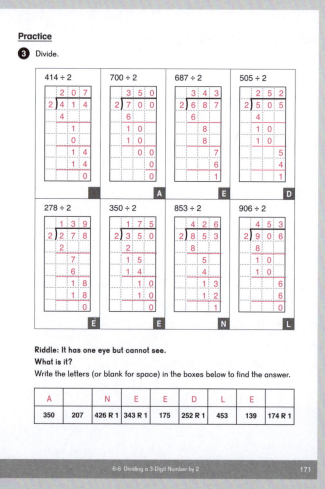

414 ÷ 2 — **L**
700 ÷ 2 — **A**
687 ÷ 2 — **E**
505 ÷ 2 — **D**
278 ÷ 2 — **E**
350 ÷ 2 — **E**
853 ÷ 2 — **N**
906 ÷ 2 — **L**

Riddle: It has one eye but cannot see.
What is it?
Write the letters (or blank for space) in the boxes below to find the answer.

A	N	E	E	D	L	E
350	207	426 R 1	343 R 1	175	252 R 1	453	139	174 R 1

---

4 Color the spaces that contain an even number.

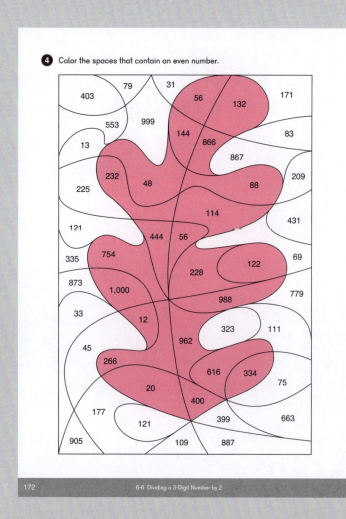

---

5 A store sold twice as many burritos on Saturday than it sold on Friday.
It sold 318 burritos on Saturday.
How many burritos did the store sell on Friday?
318 ÷ 2 = 159
The store sold 159 burritos on Friday.

6 Divide 256 by 2.
Divide the quotient again by 2.
Continue until you get a quotient of 2.
How many times did you divide by 2?
256 ÷ 2 = 128
128 ÷ 2 = 64
64 ÷ 2 = 32
32 ÷ 2 = 16
16 ÷ 2 = 8
8 ÷ 2 = 4
4 ÷ 2 = 2

7 times

Exercise 7

Basics

1 Fill in the missing digits and numbers.

(a) Divide 753 by 4.

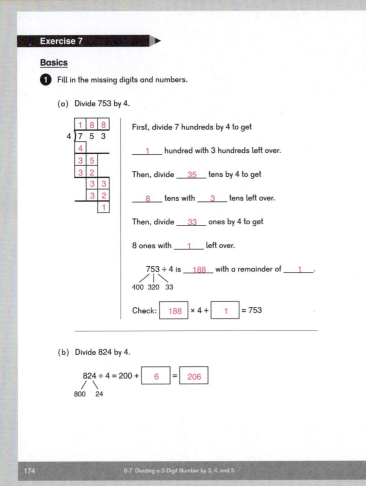

First, divide 7 hundreds by 4 to get

___1___ hundred with 3 hundreds left over.

Then, divide ___35___ tens by 4 to get

___8___ tens with ___3___ tens left over.

Then, divide ___33___ ones by 4 to get

8 ones with ___1___ left over.

753 ÷ 4 is ___188___ with a remainder of ___1___.

400 320 33

Check: [188] × 4 + [1] = 753

(b) Divide 824 by 4.

824 ÷ 4 = 200 + [6] = [206]

800 24

Practice

2 Divide.

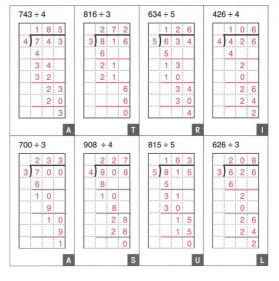

What is the only continent that does not have an active volcano?
Write the letters in the boxes below to find the answer.

A	U	S	T	R	A	L	I	A
233 R 1	163	227	272	126 R 4	185 R 3	208 R 2	106 R 2	233 R 1

3 Gavin collected $935 selling raffle tickets for a fund raiser.
Each raffle ticket cost $5.
How many raffle tickets did he sell?
935 ÷ 5 = 187
He sold 187 raffle tickets.

4 Divide 256 by 4.
Divide the quotient again by 4.
Continue until you get a quotient of 4.
How many times did you divide by 4?
256 ÷ 4 = 64
64 ÷ 4 = 16
16 ÷ 4 = 4

3 times

Challenge

5 Write the missing digits.

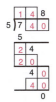 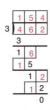

Exercise 8

Basics

1 Fill in the missing digits and numbers.

(a) Divide 468 by 5.

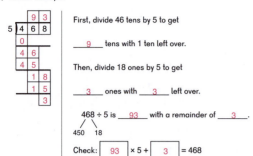

First, divide 46 tens by 5 to get

___9___ tens with 1 ten left over.

Then, divide 18 ones by 5 to get

___3___ ones with ___3___ left over.

468 ÷ 5 is ___93___ with a remainder of ___3___.

450   18

Check: 93 × 5 + 3 = 468

(b) Divide 340 by 4.

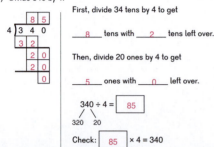

First, divide 34 tens by 4 to get

___8___ tens with ___2___ tens left over.

Then, divide 20 ones by 4 to get

___5___ ones with ___0___ left over.

340 ÷ 4 = 85

320   20

Check: 85 × 4 = 340

Practice

2 Divide.

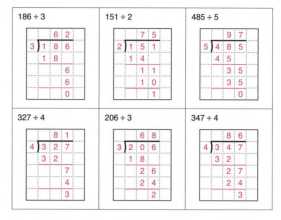

Put the quotients you found above in order from least to greatest.
Find the sum of the middle two numbers.
The answer should be 156.

62, 68, 75, 81, 86, 97

75 + 81 = 156

3 Circle the expressions where the quotient will be a 2-digit number.

443 ÷ 2   (309 ÷ 4)   (287 ÷ 3)   638 ÷ 5

743 ÷ 3   500 ÷ 2   (487 ÷ 5)   (333 ÷ 4)

4 An orchard has 255 apple trees.
It has 3 times as many apple trees as pear trees.
How many pear trees does the orchard have?

255 ÷ 3 = 85
The orchard has 85 pear trees.

5 Alex has 59 blue craft sticks, 68 red craft sticks, and 32 yellow craft sticks.
It takes 4 craft sticks to make a picture frame.
How many craft sticks will be left over after he makes as many picture frames as he can?

59 + 68 + 32 = 159
159 ÷ 4 is 39 R 3
He will have 3 craft sticks left over.

Challenge

6 Write the missing digits.

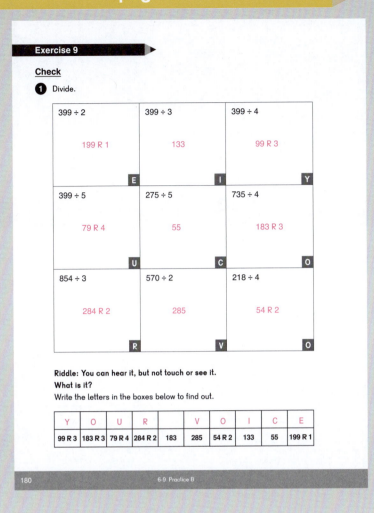

Exercise 9

Check

1 Divide.

399 ÷ 2	399 ÷ 3	399 ÷ 4
199 R 1	133	99 R 3
E	**I**	**Y**

399 ÷ 5	275 ÷ 5	735 ÷ 4
79 R 4	55	183 R 3
U	**C**	**O**

854 ÷ 3	570 ÷ 2	218 ÷ 4
284 R 2	285	54 R 2
R	**V**	**O**

Riddle: You can hear it, but not touch or see it.
What is it?
Write the letters in the boxes below to find out.

Y	O	U	R		V	O	I	C	E
99 R 3	183 R 3	79 R 4	284 R 2	183	285	54 R 2	133	55	199 R 1

180   6-9 Practice B

2 Sasha is decorating cookies with candy stars.
Each cookie gets 5 stars.
She used 690 stars.
How many cookies did she decorate?
690 ÷ 5 = 138
She decorated 138 cookies.

3 Tiara is decorating 82 round cookies and some heart-shaped cookies.
Each cookie gets 3 jelly beans.
She uses 456 jelly beans.
How many heart-shaped cookies did she decorate?
456 ÷ 3 = 152
152 − 82 = 70
She decorated 70 heart-shaped cookies.

4 Ryan is decorating cookies with gum drops and cinnamon hearts.
Each cookie gets 3 gum drops and 5 cinnamon hearts.
He has 160 gum drops and 240 cinnamon hearts.
What is the greatest number of cookies he can decorate?
160 ÷ 3 is 53 R 1
240 ÷ 5 = 48
The greatest number of cookies he can decorate is 48.

6-9 Practice B   181

Challenge

5 A jar had 5 times as many blue beads as red beads.
After 204 blue beads were used for an art project,
there were twice as many blue beads as red beads.
How many blue beads are left?

Blue
Red

204 ÷ 3 = 68
68 × 2 = 136
There are 136 blue beads left.

6 There are some row boats.
The boats are either doubles with 2 oars, or quads with 4 oars.
There are 260 oars and 100 row boats.
How many doubles and how many quads are there?
Hint: If they were all doubles, how many oars are there?

100 × 2 = 200
200 oars used so far, each boat has 2 oars.
260 − 200 = 60
60 oars left, put 2 in some boats.
60 ÷ 2 = 30
30 boats with 2 more, or 4 oars
100 − 30 = 70
70 doubles and 30 quads.

Check:
70 × 2 = 140
30 × 4 = 120
140 + 120 = 260 ✓

If students struggle, have them do the problem as if 26 oars and 10 boats, and draw pictures of the boats, putting 2 oars in each first, then 2 more with the left over oars. Point out that not all problems are best solved with bar models. If using a bar model is not helping, try a different approach.

182   6-9 Practice B

Notes

Suggested number of class periods: 4–5

	Lesson	Page	Resources		Objectives
	Chapter Opener	p. 245	TB:	p. 201	Investigate graphs and tables.
1	Picture Graphs and Bar Graphs	p. 246	TB: WB:	p. 202 p. 183	Create and interpret scaled picture graphs. Interpret bar graphs.
2	Bar Graphs and Tables	p. 249	TB: WB:	p. 207 p. 188	Create and interpret scaled bar graphs from tables.
3	Practice	p. 253	TB: WB:	p. 213 p. 192	Practice creating and interpreting graphs.
	Review 2	p. 256	TB: WB:	p. 217 p. 196	Review content from Chapter 1 through Chapter 7.
	Workbook Solutions	p. 258			

In **Dimensions Math® 2B Chapter 14**, students learned to:

- Interpret and create a scaled picture graph.
- Interpret and create a bar graph with scale of 1.
- Understand what kind of information can be determined from the graph.

In this chapter, students will learn to:

- Interpret and create scaled picture graphs and bar graphs.
- Create bar graphs from data tables.

A graph is a pictorial representation of data that is helpful for capturing characteristics of data and making comparisons between sets of objects. Picture graphs and bar graphs show the data visually, making it easier to compare the data as a whole.

Tables present data numerically, organized in a way that makes it easy to find the actual numerical data which can then be used to calculate actual differences in values.

Encourage students to compare the benefits or drawbacks of presenting data in tables, picture graphs, or bar graphs.

Minimal supplemental activities are provided, as students should spend most of each lesson creating and interpreting the graphs. Extend the lessons by asking students to think critically about what type of data they want to collect, how to draw their graphs, and data presented in graphs. Ask:

- What's the third least liked activity? What's the second to last item? (Incorporate deeper ordinal numbers.)
- Is there another way to represent this data?

After students complete the chapter in the textbook, have them continue to practice creating and reading graphs with content from other school subjects.

Materials

- Crayons or markers
- Rulers
- Sticky notes

Blackline Masters

- Centimeter Graph Paper
- Engineering Graph Paper
- Graph
- Graph Paper
- Inch Graph Paper

Objective

- Investigate graphs and tables.

Lesson Materials

- Sticky notes

Use the **Chapter Opener** as assessment of prior knowledge of surveys and graphs.

Have students discuss page 201 and recall surveys they have taken.

Continue to Lesson 1 or extend the **Chapter Opener** to a full lesson by having students work in small groups to come up with a question and survey their classmates. Limit students to 4 or 5 categories.

Have students record results with tally marks and share with the class.

As review, choose one of the surveys and create a graph on the board from the data, using sticky notes for each tally mark.

Alternatively, groups can create a graph from their data.

Chapter 7

Graphs and Tables

The grade 3 students went to an outdoor camp. At the end of the camp, they collected information to help plan the next camping trip.

What was your favorite activity at camp?

What was your favorite campfire food?

Which was your favorite hiking route?

Which campfire song did you like best?

201

Lesson 1 Picture Graphs and Bar Graphs

Objectives

- Create and interpret scaled picture graphs.
- Interpret bar graphs.

Lesson Materials

- Graph (BLM)
- Rulers
- Crayons or markers

Think

Discuss the information recorded in Emma's tally chart in **Think**, as well as question (a). Have students share which category is easy to represent using one square for 5 responses (Science talks) and which ones cannot be represented this way. Have students discuss how they might represent the categories that are not groups of 5. Provide each student with a Graph (BLM) to complete task (a) with crayons or markers, then continue with **Think** questions.

When students have finished their graphs, have them think about questions (b), (c), and (d), and how a graph can make it easier to find answers to those questions.

Learn

Have students compare their graphs to the one in **Learn**. Discuss how Emma uses partial blocks to represent quantities less than 5. Dion points out that the graph makes it easier to see differences in the numbers of students who chose each activity.

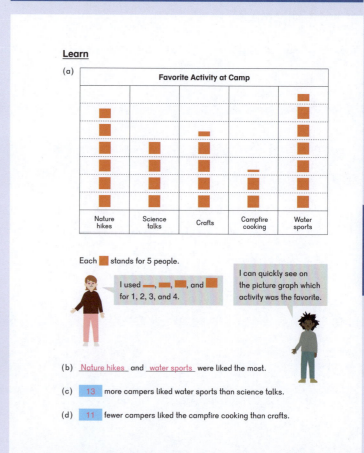

Lesson 1
Picture Graphs and Bar Graphs ①

Think

Emma **surveyed** campers to find out which camp activity they liked the best. She recorded her **data** with tally marks.

Activity	Tally	Number
Nature hikes	卌 卌 卌 卌 卌 ////	29
Science talks	卌 卌 卌 卌	20
Crafts	卌 卌 卌 卌 //	22
Campfire cooking	卌 卌 /	11
Water sports	卌 卌 卌 卌 卌 卌 ///	33

(a) Display the information in a graph. Use one ▮ to show 5 students.

> How can we show 1, 2, 3, or 4 students?

(b) Which two activities were liked the most?

(c) How many more campers liked water sports than science talks?

(d) How many fewer campers liked the campfire cooking than crafts?

202   7-1 Picture Graphs and Bar Graphs

Learn

(a)

Favorite Activity at Camp

| Nature hikes | Science talks | Crafts | Campfire cooking | Water sports |

Each ▮ stands for 5 people.

> I used ▬, ▬, ▮, and ▮ for 1, 2, 3, and 4.

> I can quickly see on the picture graph which activity was the favorite.

(b) _Nature hikes_ and _water sports_ were liked the most.

(c) _13_ more campers liked water sports than science talks.

(d) _11_ fewer campers liked the campfire cooking than crafts.

7-1 Picture Graphs and Bar Graphs   203

Have students compare and discuss the similarities and differences between the picture graph and the bar graph in the textbook. They should note:

- The scale on the left side of the graph is marked in increments of 5 campers.
- The scale on the left is labeled "Number of Campers."
- Activities with quantities that aren't a multiple of 5 fill only part of a rectangle.

Discuss Mei and Sofia's comments. Sofia points out that a graph presents a visual representation of the data.

Note: Students will create a bar graph in the next lesson.

Do

❶ Students may need to record the total number of each badge awarded, similar to the answer overlay.

(f) Students should add the total number of each type of badge awarded:

7 + 17 + 7 + 11 + 2 + 8 + 14 + 9 = 75 badges in all.

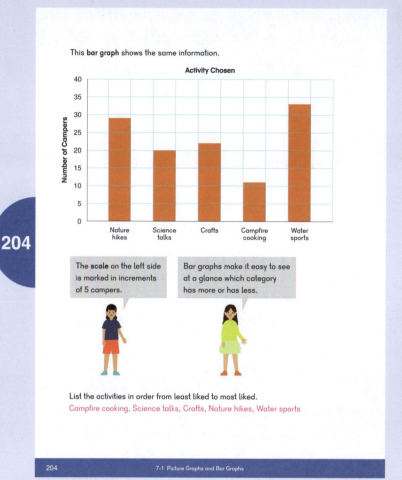

204

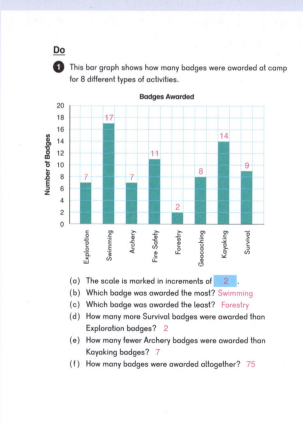

205

② Have students compare and discuss the differences between the bar graphs in ① and ②. They may note:

- This graph in ② goes across (is horizontal).
- The scale is above the information in ②. It is labeled "Number of Campers" and numbered by tens.
- On the horizontal graph (②), the longer the bar the greater the number, and on the vertical graph (①), the higher the bar the greater the number.
- There are smaller tick marks between the increments of 10 on the graph in ②. On this graph you can determine the exact value for each category by looking at the scale.
  - To help students understand the smaller increments, have them compare the tick marks and increments on a ruler or a number line (see Chapter 1) to the ones on the graph.
- In ②, the category bars look like the division bar models. To determine the increments of the blue boxes, students could see that the Number of Campers who chose Meadow totals 15, and there are 3 squares to represent the quantity, so the value of each square is 5.

③ This question points out the importance of reading the scale when interpreting data.

Exercise 1 • page 183

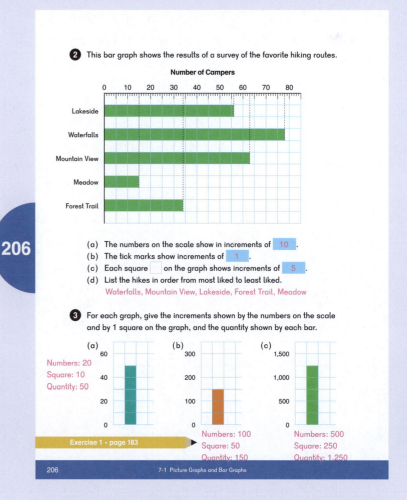

② This bar graph shows the results of a survey of the favorite hiking routes.

Number of Campers

(a) The numbers on the scale show in increments of [10].
(b) The tick marks show increments of [1].
(c) Each square ☐ on the graph shows increments of [5].
(d) List the hikes in order from most liked to least liked.
Waterfalls, Mountain View, Lakeside, Forest Trail, Meadow

③ For each graph, give the increments shown by the numbers on the scale and by 1 square on the graph, and the quantity shown by each bar.

(a) Numbers: 20 Square: 10 Quantity: 50

(b) Numbers: 100 Square: 50 Quantity: 150

(c) Numbers: 500 Square: 250 Quantity: 1,250

Exercise 1 • page 183

206

Objective

- Create and interpret scaled bar graphs from tables.

Lesson Materials

- Graph Paper (BLM)
- Crayons or markers

Think

Provide students with Graph Paper (BLM) and crayons. Pose the **Think** task. Discuss the differences between the table and a graph, and which will present a clearer representation of the data.

Discuss step 1, if needed, prior to having students create their graphs. Prompt students to consider what increments might be helpful to them. They should note that the quantities range from 15 to 92, so their graph should be able to represent data within 100. Most students will likely choose scales that are multiples of 2, 5, or 10.

Have students create a graph for the data in the table on textbook page 207. Creating the graph will take the majority of the lesson time.

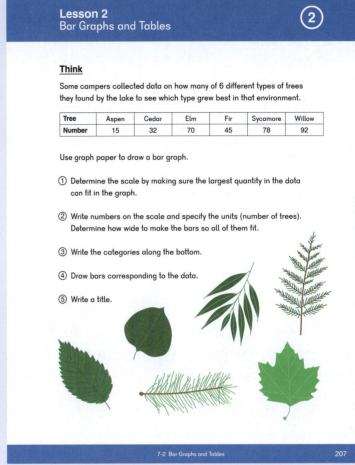

Lesson 2
Bar Graphs and Tables ②

Think

Some campers collected data on how many of 6 different types of trees they found by the lake to see which type grew best in that environment.

Tree	Aspen	Cedar	Elm	Fir	Sycamore	Willow
Number	15	32	70	45	78	92

Use graph paper to draw a bar graph.

① Determine the scale by making sure the largest quantity in the data can fit in the graph.

② Write numbers on the scale and specify the units (number of trees). Determine how wide to make the bars so all of them fit.

③ Write the categories along the bottom.

④ Draw bars corresponding to the data.

⑤ Write a title.

7-2 Bar Graphs and Tables   207

207

Learn

Have students compare their graphs to Dion's graph. Discuss the important features of the graph (title, axes labels, scale).

Discuss which representation gives students a more accurate picture of the data.

Ask students, "What are the advantages and disadvantages of using a table? A graph?"

Ask students why Dion made his bars two boxes wide. (Answers will vary. Examples: To accommodate the longer tree names, to use the full page.)

Ask students if they used one box or more than one box for the width of their bars. Show a graph with one-box width and ask students if the answers from this graph are the same as the answers from Dion's graph.

Remind students that graphs are a pictorial representation of data and should be easy to read and understand. The width of the bars in this graph does not matter, the height shows the quantities.

Do

❷ This question shows that the types of trees do not need to be written in the order they were given on the table, but can be rearranged to show the size of the data in each category from greatest to least or least to greatest.

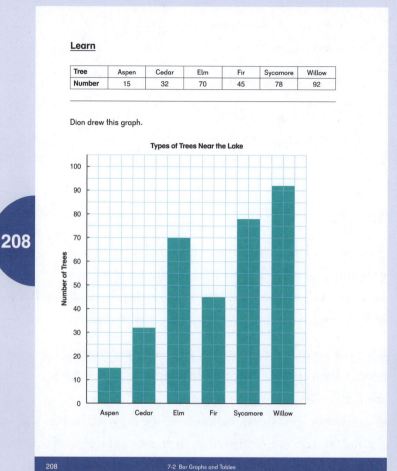

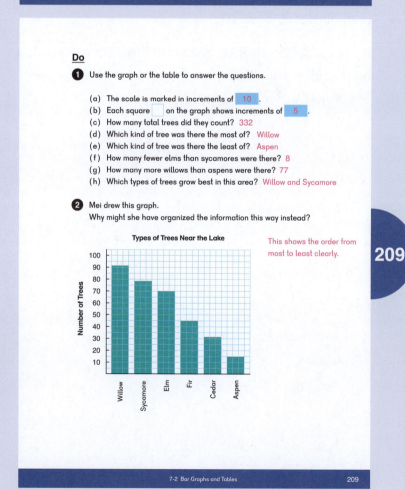

208

209

3 Discuss Alex's comments about the "other" category.

4 Use the table and graph from **3** to answer the questions. Have students share which they used to answer the questions, the table or the graph.

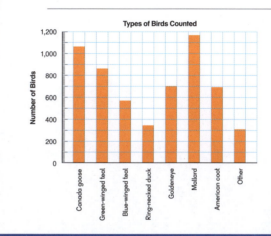

3 Some scientists gave a presentation about bird migration in some wetlands near the camp. The table and graph show the number of waterfowl counted over a certain period of time.

Bird	Number
Canada goose	1,062
Gadwall	40
Green-winged teal	865
Blue-winged teal	582
Northern shoveler	189
Northern pintail	52
Ring-necked duck	341
Goldeneye	698
Mallard	1,172
Canvasback	31
American coot	692

Types of Birds Counted

In the graph, the category "other" was created to include birds that were in small numbers. Which birds are shown in this category on the graph?

Gadwall, Northern shoveler, Northern pintail, Canvasback

The category Other is often put at the end.

4 Answer the following questions for the birds. Did you use the graph, the table, or both for each? Answers may vary.

(a) Approximately how many birds were counted for this data?
About 5,500

(b) Which bird was seen the most?
Mallard

(c) List the types of birds in order from least to greatest number.
Canvasback, Gadwall, Northern pintail, Northern shoveler, Ring-necked duck, Blue-winged teal, American Coot, Goldeneye, Green-winged teal, Canada goose, Mallard

(d) How many teals were seen in all?
1,447

(e) For which types were there more than 600 birds counted?
American Coot, Goldeneye, Green-winged teal, Canada goose, Mallard

(f) For which types were there less than 600 birds counted?
Canvasback, Gadwall, Northern pintail, Northern shoveler, Ring-necked duck, Blue-winged teal

(g) There were a little over 300 of which type of bird?
Ring-necked Duck

5 All three graphs have the same quantities and data. They are scaled differently. Note that people use different scales to emphasize something.

Extend (d) by looking at graphs in advertising and comparing the visual representations used. What axes and scales do the different graphs use?

Exercise 2 • page 188

212

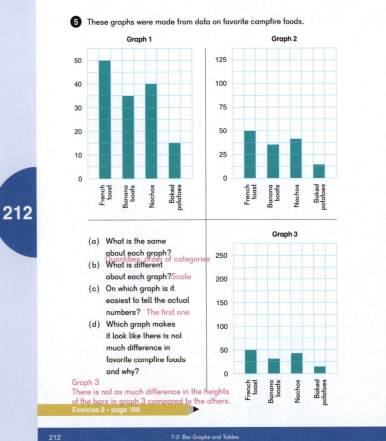

Teacher's Guide 3A Chapter 7 © 2017 Singapore Math Inc.

Lesson 3 Practice

Objective

- Practice creating and interpreting graphs.

After students complete the **Practice** in the textbook, have them continue to practice interpreting data from graphs with content from other school subjects.

1 This is a survey of some water sports that campers would like to do. Campers could vote for up to two activities.

Activity	Number
Kayak	61
Canoe	35
Sail	59
Snorkel	19
Wakeboard	73

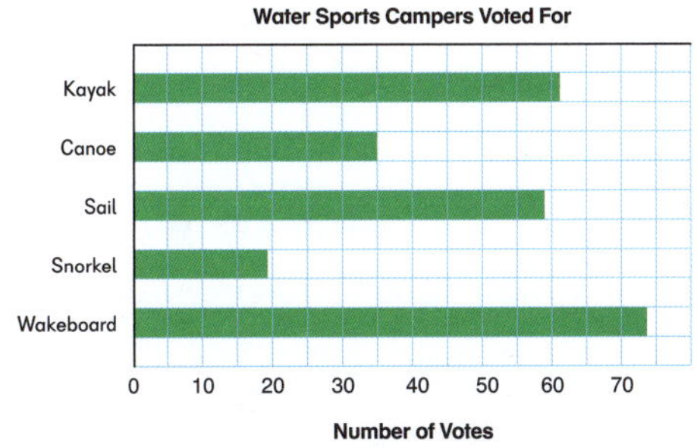

Water Sports Campers Voted For

Number of Votes

(a) What increment is represented by 1 square on the graph? 5
(b) List the activities in order from least to greatest number of times voted for.
(c) Which two activities were voted for almost the same number of times?
(d) What is the difference in number of votes between the most popular and the least popular activity?

(b) Snorkel, Canoe, Sail, Kayak, Wakeboard
(c) Kayak and Sail
(d) 54 votes

213

2 Tell students the graph is called a Double Bar Graph. Have them discuss what makes it a "double."

Discuss the table and double bar graph with students. Suggested questions to ask:

- Which gives a clearer picture of the data, the table or the graph?
- How is this table different from previous tables in the chapter?
- What information does the bar graph show?
- How is this graph different from previous bar graphs in this chapter?
- Why might this graph be drawn this way?
- Can you think of other data that might be shown with a double bar graph?

214

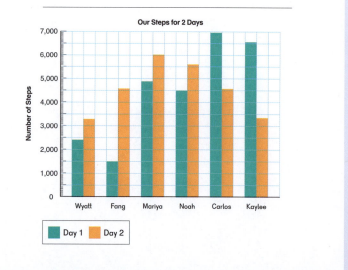

2 Some students wore pedometers and recorded the total number of steps they took each day for two days. They drew a graph to show the results.

Name	Day 1	Day 2	Total	Difference
Wyatt	2,416	3,290	5,705	874
Fang	1,504	4,578	6,082	3,074
Mariya	4,892	6,009	10,901	1,117
Noah	4,509	5,605	10,114	1,096
Carlos	6,953	4,570	11,523	2,383
Kaylee	6,561	3,350	9,911	3,211

(a) Each square on the graph shows an increment of   500 .
(b) Who recorded more steps on Day 1 than Day 2?   Carlos and Kaylee
(c) Who recorded more steps on Day 2 than Day 1?   Wyatt, Fang,
(d) Who recorded the most steps for both days combined?   Mariya, and Noah
 Use estimation.   Carlos
(e) Who recorded the fewest steps for both days combined?
 Use estimation.   Wyatt
(f) Who had the greatest difference in number of steps between the two days?   Kaylee
(g) Who had the least difference in number of steps between the two days?   Wyatt
(h) Wyatt (5,705), Fang (6,082), Kaylee (9,911) recorded for both days
 Noah (10,114), Mariya (10,901), Carlos (11,523)
(i) List the difference in number of steps each student recorded for both days in order from least to greatest.
 Wyatt (874), Noah (1,096),
 Mariya (1,117), Carlos (2,383), Fang (3,074), Kaylee (3,211)

215

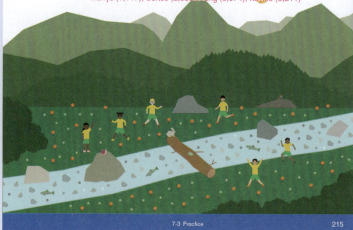

3 Have students imagine that each time an injury occurred at camp, the type of injury was recorded. Ask, "Why isn't this first chart helpful?" (It's hard to count and see how many of each kind of injury there were.)

Have students make a tally chart and then copy it into the table for (b).

Activity

▲ **Graph It**

Materials: Centimeter Graph paper (BLM), Inch Graph Paper (BLM), Engineering Graph Paper (BLM), crayons or markers

Have students use the different styles of graph paper to graph data from the lessons.

Students can compare which style of graph paper displays the data the most clearly.

Exercise 3 • page 192

216

3 (a) Use the chart to tally each type of injury at camp.

Scrape: 24   Rash: 12
Cut: 21   Burn: 7
Bug bite: 16   Bee sting: 3
Bruise: 12   Sprain: 3

Bruise	Bug Bite	Burn	Scrape	Bug Bite	Cut	Bruise
Burn	Cut	Rash	Cut	Rash	Scrape	Scrape
Scrape	Scrape	Scrape	Bug Bite	Bug Bite	Sprain	Scrape
Bruise	Bruise	Cut	Bug Bite	Cut	Bug Bite	Cut
Rash	Bug Bite	Burn	Bruise	Bug Bite	Bug Bite	Scrape
Bug Bite	Bruise	Cut	Scrape	Rash	Bruise	Rash
Cut	Rash	Rash	Scrape	Cut	Bruise	Bruise
Bee Sting	Burn	Bug Bite	Sprain	Rash	Scrape	Scrape
Scrape	Scrape	Cut	Burn	Scrape	Bug Bite	Cut
Rash	Cut	Scrape	Scrape	Bruise	Bruise	Cut
Scrape	Scrape	Cut	Bee Sting	Cut	Sprain	Scrape
Cut	Rash	Rash	Bug Bite	Cut	Bruise	Bee Sting
Cut	Burn	Scrape	Scrape	Bug Bite	Cut	Cut
Bug Bite	Cut	Rash	Burn	Bug Bite	Scrape	Scrape

(b) Copy and complete this table from the data.

Injury	Bruise	Cut	Rash	Scrape	Bug Bite	Other
Number	12	21	12	24	16	13

(c) Draw a bar graph for this data.

(d) Which was the most common injury?   Scrape

(e) List the five next most common injuries in order of most common to least common. Scrape, cut, bug bite, rash, bruise
Note: There are the same number of rashes and bruises

(f) How might this data help camp counselors?
Answers will vary. They could stock more supplies for scrapes and cuts, more bug spray, and teach campers how to avoid the most common injuries.

Exercise 3 • page 192

216   7-3 Practice

Review 2

Objective

- Review content from Chapter 1 through Chapter 7.

Use these pages to review content before doing a mid-year cumulative assessment. This review can also serve as the basis for assessment.

Reviews are important for reinforcing skills. They also provide teachers with information regarding which students may need remediation and specifically which topics should be included.

217

❶ Estimate, then find the value.   Estimations may vary.

(a) 4,890 + 283   (b) 6,785 + 2,295
   about 5,200; 5,173   about 9,000; 9,080
(c) 9,125 − 864   (d) 9,006 − 4,237
   about 8,110; 8,261   about 5,000; 4,769

❷ Find the value.

(a) 65 × 4 260   (b) 70 × 0 0   (c) 98 × 3 294

(d) 849 × 2 1,698   (e) 307 × 9 2,763 (f) 474 ÷ 5 94 R 4

(g) 67 ÷ 3 22 R 1 (h) 59 ÷ 4 14 R 3 (i) 963 ÷ 2 481 R 1

(j) 85 ÷ 1 85   (k) 85 ÷ 85 1   (l) 0 ÷ 85 0

❸ (a) 39 ÷ 4 is 9 with a remainder of 3.

  (b) 145 ÷ 3 is 48 with a remainder of 1.

❹ (a) What is the least odd number that can be formed using the digits 7, 8, 0, and 3?
      3,087
  (b) What is the greatest even number that can be formed using the digits 6, 3, 5, and 9?
      9,536

218

❺ This table and graph below shows the number of campsites rented at a park each night for one week.

On Friday, Saturday, and Sunday the cost for renting a campsite is $5 a night.

On Monday through Thursday, the cost for renting a campsite is $4 a night.

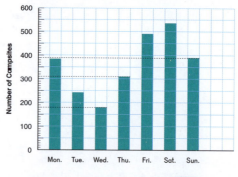

(a) The scale is numbered in increments of 100 .

(b) Each tick mark shows an increment of 10 .

(c) Each square on the graph is for an increment of 50 .

Brain Works

★ **Number Patterns**

(a) $1 + 2 + 3 + 2 + 1 =$ ▢

(b) $1 + 2 + 3 + 4 + 5 + 6 + 5 + 4 + 3 + 2 + 1 =$ ▢

(c) $1 + 2 + 3 + 4 + 5 + 6 + 7 + 6 + 5 + 4 + 3 + 2 + 1$ $=$ ▢

(d) $1 + 2 + 3 + 4 + 5 + 6 + 7 + 8 + 9 + 8 + 7 + 6 + 5$ $+ 4 + 3 + 2 + 1 =$ ▢

(e) $1 + 2 + 3 + ... + 99 + 100 + 99 + ... + 3 + 2 + 1$ $=$ ▢

Answers

(a) 9

(b) 36

(c) 49

(d) 81

(e) 10,000

(d) On which day of the week were the greatest number of campsites rented? Saturday

(e) On which day of the week were the fewest number of campsites rented? Wednesday

(f) Which two days had about the same number of campsites rented?
Monday and Sunday

(g) Complete the table.

Day	Mon.	Tue.	Wed.	Thu.	Fri.	Sat.	Sun.
Number	382	241	180	310	493	532	390

(h) How many more campsites were rented on Saturday than on Sunday?
532 − 390 = 142; 142 more

(i) Estimate how many campsites were rented all week by rounding the daily rental to the nearest 100.   2,500

(j) How much money did the camp receive on Friday?
493 × 5 = 2,465; $2,465

(k) How much money did the camp receive Tuesday through Thursday?
241 + 180 + 310 = 731; 731 × 4 = 2,924

(l) How much more money did the camp receive on Saturday than on Wednesday?   180 × 4 = 720; 532 × 5 = 2,660
2,660 − 720 = 1,940; $1,940

(m) Some sites were rented for a single night and others for multiple nights. If the camp received $905 for single night rentals on Friday, how many sites were rented for multiple nights that night?   905 ÷ 5 = 181
493 − 181 = 312
312 sites were rented for multiple nights.

219

---

6 There are 540 angelfish in a pet store.
There are 5 times as many goldfish as angelfish.
How many fewer angelfish are there than goldfish?
540 × 4 = 2,160   or: 540 × 5 = 2,700
2,160 fewer angelfish   2,700 − 540 = 2,160

goldfish
angelfish
540
?

7 A florist had 285 roses.
19 of them wilted.
She wants to make bouquets of 3 roses each with the remaining roses.
How many bouquets can she make? 285 − 19 = 266
266 ÷ 3 is 88 R 2
88 bouquets

8 A teacher bought 32 notebooks at $3 each.
He paid for the notebooks and got $4 change.
How much money did he give the cashier? 32 × 3 = 96
96 + 4 = 100
$100

$3.00
NOTE BOOK
198 pages / 3" × 5"
Recycled Materials

9 A roll of ribbon 315 ft long is cut into 3 pieces, A, B, and C.   315 ft + 25 ft = 340 ft
B is 25 ft longer than A.   340 ft ÷ 4 = 85 ft
C is twice as long as B.   85 ft − 25 ft = 60 ft
How long is A?   60 ft long

C
B   315
A
25

10 8 lamp posts are an equal distance from each other along a street.
The distance from the 2nd to the 6th lamp post is 500 ft.
What is the distance from the 3rd to the 8th lamp post?
There are 4 gaps between 2nd and 6th: 500 ft ÷ 4 = 125 ft
There are 5 gaps between 3rd and 8th: 125 ft × 5 = 625 ft
625 ft

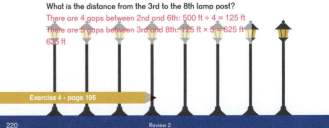

220

Chapter 7 Graphs and Tables

Exercise 1

Basics

1. Look at the graphs shown here.

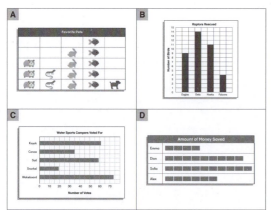

(a) Which are picture graphs? **A, D**

(b) Which are bar graphs? **B, C**

(c) Which type of graph has a numerical scale on one side of the graph?

Bar graph

(d) How many categories are there in graph C? **5**

Practice

2. This picture graph shows the number of children that signed up for different Discovery Camps at the Community Center last summer.

Discovery Camps

| Ooey Gooey | Jurassic Journey | Under the Sea | Treasure Island | Inventors Workshop |

Each ● stands for 4 children.

(a) List the summer camps in order from most popular to least popular.

Jurassic Journey, Inventors Workshop, Treasure Island, Under the Sea, Ooey Gooey

(b) How many more children signed up for Inventor's Workshop than for Treasure Island? **9**

(c) How many fewer children signed up for Ooey Gooey than for Jurassic Journey? **21**

(d) Complete this bar graph with the information from the picture graph.

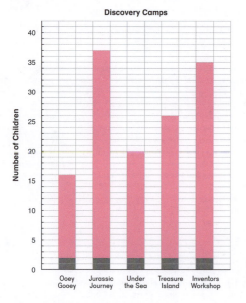

Discovery Camps

(e) The scale is numbered in increments of **5**

(f) Each tick mark on the graph shows an increment of **1**.

(g) On which type of graph is it easier to read the numbers for each category?

Bar Graph

3. This bar graph shows the number of each kind of shoe sold by a sports store over a period of time.

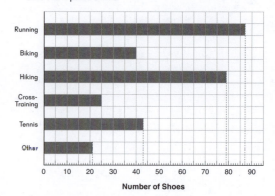

Number of Shoes

(a) The scale is numbered in increments of **10**.

(b) Each square ☐ on the graph shows increments of **5**.

(c) Each tick mark on the graph shows increments of **1**.

(d) Which type of shoe did they sell the most of? **Running**

(e) Which two types of shoes did they sell almost the same number of?

Biking and Tennis

(f) The store also sells water sport shoes and golf shoes. Under which category are these shoes graphed?

Other

(g) Use the information from the graph to complete the table.

Shoe	Running	Biking	Hiking	Cross-training	Tennis	Other
Number	87	40	79	25	43	21

(h) How many more running shoes than tennis shoes were sold? 44

(i) How many fewer biking shoes than hiking shoes were sold? 39

4 These two graphs were created to show the results from counting the number of each type of vehicle that parked at two parking garages.

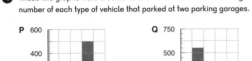

(a) Which garage had more cars? Q

(b) How many trucks were in each parking garage?

P 300

Q 250

Exercise 2

Basics

1 Chapa surveyed some students to find out which musical instrument they wanted to learn how to play.

Instrument	Number
Piano	63
Violin	75
Oboe	21
Flute	53
Trumpet	25

(a) Complete the bar graph below with this information.
Order of categories may vary.

Instruments Students Want to Learn to Play

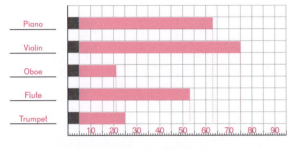

Number of Students

Students are not required to draw the dotted lines, but they can use a ruler to determine where the bars should end.

(b) List the instruments in order from most popular to least popular.

Violin, Piano, Flute, Trumpet, Oboe

(c) Graph the same information on the two graphs below.

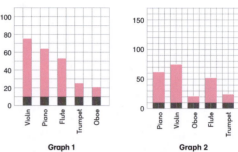

Graph 1   **Graph 2**

(d) What differences do you notice between the two graphs?
Answers will vary. Examples:
The scales are different. The scale on the graph on the left has smaller increments.
The categories are in a different order between the two graphs.

Practice

2 Andrew wanted to find out which color is the most popular for cars. He tallied the number of cars he saw of different colors passing in front of his apartment building in an hour, and recorded his information in a table.

Color	Number
White	57
Black	60
Blue	48
Gray/Silver	87
Red	30
Brown/Beige	3
Other	15

(a) Create a bar graph for this information on the next page.

(b) Under which category did Andrew put green cars? _____Other_____

(c) Under which category would he put cars that are more than one color?
_____Other_____

(d) Under which category did he likely put tan cars? _____Brown/Beige_____

(e) Which car color is most popular? _____Gray/Silver_____

(f) Which are the three most popular colors? Gray/Silver, Black, White

Graphs may vary.

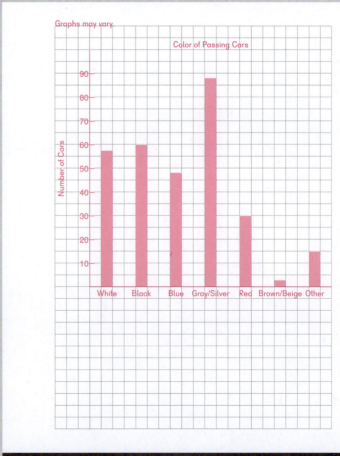

Exercise 3

Check

1 Heather wanted to find out which type of bird seed the birds that came to her backyard preferred.

She put the birdseeds in different bird feeders and then counted the number of birds that came to each feeder for a period of time.

Birdseed	Sunflower	Millet	Sorghum	Corn
Number of birds	36	54	28	17

(a) Create a bar graph for this information.

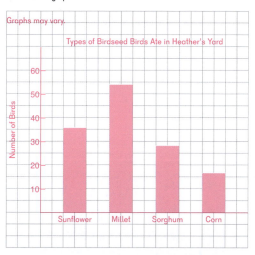

(b) List the birdseed types from most popular to least popular.
Millet, Sunflower, Sorghum, Corn

(c) How many more birds came to eat the millet than the corn? ⬚ 37

(d) Is there another way she could collect data to find out which type of seed was most popular that would be easier than counting birds?
She could weigh or measure the amount of the birdseed before and after a period of time.

2 Create a picture graph in the space below for Heather's data.
Decide what kind of symbol to use, whether it should stand for 2, 3, 4, or 5 birds, and how to represent numbers that do not divide evenly.

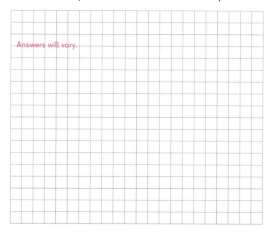

Answers will vary.

3 This graph shows the number of four kinds of salmon that were migrating up a stream on the first day of three different months.

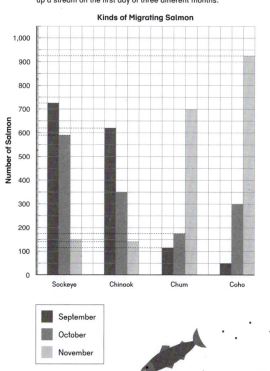

(a) Which kind of salmon migrate primarily in September and October?
Sockeye, Chinook

(b) Which kind of salmon migrate primarily in November?
Chum, Coho

(c) Complete the table using the information from the graph.

	September	October	November
Sockeye	725	590	150
Chinook	620	350	140
Chum	115	175	700
Coho	50	300	925

(d) List each kind of salmon in order from least to greatest number counted. (Use estimation)
Sockeye, Coho, Chinook, Chum

(e) How many more Coho were counted in November than in September and October combined?
300 + 50 = 350
925 − 350 = 575

(f) How many fewer Chinook were counted in November than in September and October combined?
620 + 350 = 970
970 − 140 = 830

Exercise 4

Check

1 Find the value.

897 + 219	806 + 7,496	6,399 + 2,402
897 + 219 1,116	7,496 + 806 8,302	6,399 + 2,402 8,801
3,290 − 524	9,751 − 5,438	7,006 − 1,528
3,290 − 524 2,766	9,751 − 5,438 4,313	7,006 − 1,528 5,478
67 × 3	570 × 2	4 × 864
67 × 3 201	570 × 2 1,140	864 × 4 3,456
58 ÷ 3	570 ÷ 4	218 ÷ 5
19 3)58 3 28 27 1	142 4)570 4 17 16 10 8 2	43 5)218 20 18 15 3

2 Write >, <, or = in the ◯.

(a) 7,632 ⟨>⟩ 700 + 6,000 + 30 + 2

(b) 400 + 5,000 + 90 ⟨<⟩ 5,100 + 490

(c) 320 tens ⟨>⟩ 20 hundreds + 100 tens

(d) 4,968 + 3,125 ⟨>⟩ 9,207 − 3,895

(e) 385 + 543 + 220 + 50 ⟨<⟩ 498 + 420 + 90 + 487

(f) 444 × 5 ⟨=⟩ 555 × 4

3 Write the number word for 48 tens and 6 ones.
four hundred eighty-six

4 Use mental calculation to add or subtract in order.

360 + 80 − 70 + 98 − 68 = ☐ 400

5 Round 8,285...

(a) To the nearest thousand.   ☐ 8,000

(b) To the nearest hundred.   ☐ 8,300

(c) To the nearest ten.   ☐ 8,290

6 Katia estimates 5,688 + 2,042 by calculating 6,000 + 2,000.
Will the estimated answer be greater or less than the actual answer?

The estimated answer will be greater than the actual answer, since
5,688 is farther from 6,000 than 2,042 is from 2,000.

7 Circle the values that can be evenly divided by 2.

(235 + 347)	172 + 621	(82 + 788)
47 × 5	(44 × 3)	(236 × 4)

8 Landon picked up trash one day at the park.
He kept track of how many items of each kind of trash he picked up and
made this table.

Paper	Plastic	Styrofoam	Glass	Metal	Other
55	64	82	8	15	12

(a) Complete bar graph for this information on the next page.

(b) Which kind of trash did he find the most of?
Styrofoam

(c) How many more of the three most common types of trash did he pick
up than the rest?
He picked up 166 more pieces of the three most common types of trash.

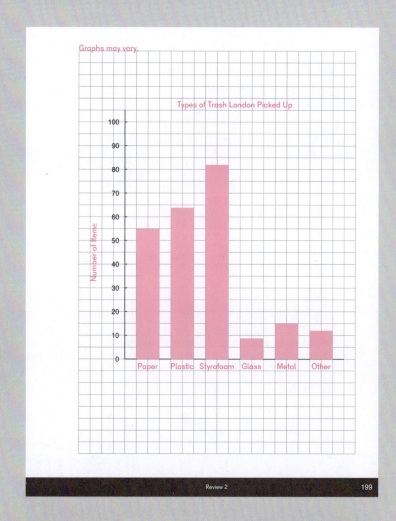

Graphs may vary.

Types of Trash Landon Picked Up

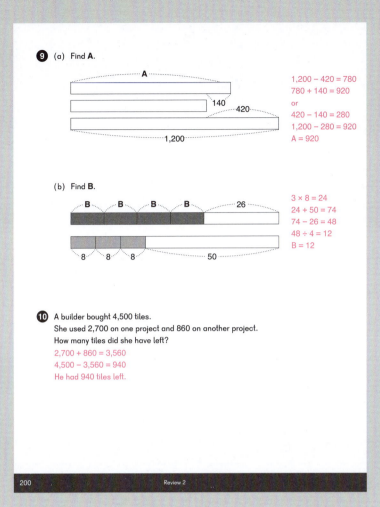

9 (a) Find **A.**

A

140

420

1,200

$1,200 - 420 = 780$
$780 + 140 = 920$
or
$420 - 140 = 280$
$1,200 - 280 = 920$
$A = 920$

(b) Find **B.**

B B B B 26

8 8 8 50

$3 \times 8 = 24$
$24 + 50 = 74$
$74 - 26 = 48$
$48 \div 4 = 12$
$B = 12$

10 A builder bought 4,500 tiles.
She used 2,700 on one project and 860 on another project.
How many tiles did she have left?
$2,700 + 860 = 3,560$
$4,500 - 3,560 = 940$
He had 940 tiles left.

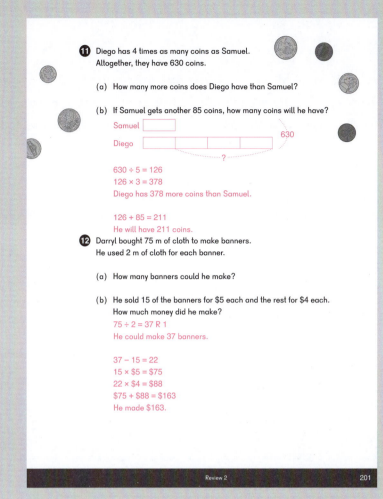

11 Diego has 4 times as many coins as Samuel.
Altogether, they have 630 coins.

(a) How many more coins does Diego have than Samuel?

(b) If Samuel gets another 85 coins, how many coins will he have?

Samuel

Diego   630

?

$630 \div 5 = 126$
$126 \times 3 = 378$
Diego has 378 more coins than Samuel.

$126 + 85 = 211$
He will have 211 coins.

12 Darryl bought 75 m of cloth to make banners.
He used 2 m of cloth for each banner.

(a) How many banners could he make?

(b) He sold 15 of the banners for $5 each and the rest for $4 each.
How much money did he make?
$75 \div 2 = 37 \text{ R } 1$
He could make 37 banners.

$37 - 15 = 22$
$15 \times \$5 = \$75$
$22 \times \$4 = \$88$
$\$75 + \$88 = \$163$
He made $163.

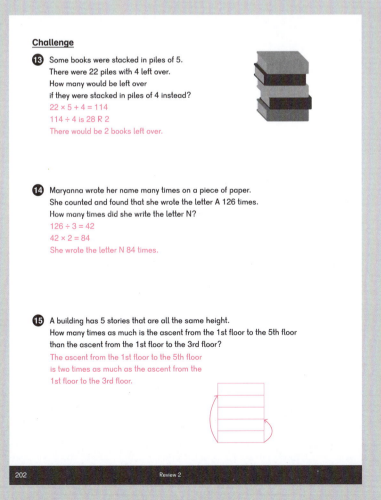

Challenge

13 Some books were stacked in piles of 5.
There were 22 piles with 4 left over.
How many would be left over
if they were stacked in piles of 4 instead?
$22 \times 5 + 4 = 114$
$114 \div 4$ is 28 R 2
There would be 2 books left over.

14 Maryanna wrote her name many times on a piece of paper.
She counted and found that she wrote the letter A 126 times.
How many times did she write the letter N?
$126 \div 3 = 42$
$42 \times 2 = 84$
She wrote the letter N 84 times.

15 A building has 5 stories that are all the same height.
How many times as much is the ascent from the 1st floor to the 5th floor
than the ascent from the 1st floor to the 3rd floor?
The ascent from the 1st floor to the 5th floor
is two times as much as the ascent from the
1st floor to the 3rd floor.

Teacher's Guide 3A Chapter 7

Blackline Masters for 3A

All Blackline Masters used in the guide can be downloaded from dimensionsmath.com.

This lists BLMs used in the **Think** and **Learn** sections.

BLMs used in **Activities** are included in the Materials list within each chapter.

Graph	**Chapter 7:** Lesson 1
Graph Paper	**Chapter 7:** Lesson 2
Inch Graph Paper	**Chapter 3:** Lesson 1
Missing Products 4-2	**Chapter 4:** Lesson 2
Number Cards	**Chapter 2:** Lesson 5
Number Line	**Chapter 1:** Lesson 5, Lesson 8, Lesson 9, Lesson 10
Place-value Cards	**Chapter 1:** Lesson 1, Lesson 2, Lesson 3, Lesson 4
Place-value Organizer	**Chapter 1:** Lesson 2 **Chapter 3:** Lesson 1, Lesson 2, Lesson 3